CLIMATE

OF UNCERTAINTY

A BALANCED LOOK AT GLOBAL WARMING AND RENEWABLE ENERGY

WILLIAM STEWART

O**C**EAN PUBLISHING

Ocean Publishing
P.O. Box 1080
200 S. Oceanshore Blvd., Suite 4
Flagler Beach, FL 32136
(386) 517-1600 • (888) 71-OCEAN
www.oceanpublishing.org

Library of Congress Cataloging-in-Publication Data

Stewart, William (William F.)
  Climate of uncertainty : a balanced look at global warming and renewable energy / William Stewart.
    p. cm.
  Includes bibliographical references and index.
  Summary: "Examines the major questions of today: global warming, renewable energy, expanding populations, and sustainability. Without taking sides, presents factual information in a clear and accessible manner" —Provided by publisher.
    ISBN 978-0-9767291-6-7 (alk. paper)
  1. Global warming. 2. Renewable energy sources. 3. Sustainability. I. Title.
    QC981.8.G56S78   2010
    363.738'74—dc22

                                                    2009049590

Cover Design: Carol Tornatore
Interior Designer/Typesetting: Gary Rosenberg
In-house Editor: Mike Cavaliere
Copyeditor: Kathleen Go

Printed in the United States of America

10   9   8   7   6   5   4   3   2   1

Contents

*To my children Melissa, Mike, Meg, Kris, CJ, Will and Luke,
to each of my fifteen bright-eyed nieces and nephews,
and to Grace, Sarah, and Jeremiah:*

*I am concerned that, despite all of our celebrated
accomplishments, my generation will deliver onto yours
a world rife with extraordinary peril and complex challenges.
I hope you will accept this book in the spirit it is intended—
my modest contribution towards a safer and
more prosperous tomorrow.*

Acknowledgements

I thank Nancy Portney for her excellent contributions in editing, challenging, drafting, guidance and support. This book would have been a far lesser enterprise without her heroic efforts.

Danielle Willard's unique blend of technical wizardry and uncommonly good judgment greatly assisted me during the months of writing.

Donna D'Emilio is simply the best assistant a person could ever wish for—and her encouragement and ongoing personal commitment to this project are deeply appreciated.

I also want to acknowledge my partners Steve Cozen, Tad Decker, Vince McGuinness, Bill Shelley, and Tom Jones of Cozen O'Connor for their unwavering support.

And, loving thanks go to my parents, step-parents, and two of my children who helped through guidance and support (you know who you are).

Most of all, thank you Diana for your inspiration, kindness, humor and sage advice. There are no words to express what you mean to me.

Preface

In November of 2008, while home from Penn State on Thanksgiving break, my daughter Megan asked me a seemingly simple question: "I think it's time I learn more about this global warming stuff—what book do you think I should read first?" Because it is a rare occasion when one of my teenagers sees fit to ask me for advice, I wanted to make the perfect recommendation. But, the more I thought about the question, the more frustrated I became.

It is not like there is an absence of terrific books on climate change, renewable energy, and sustainability. To the contrary, a number of books (all described in the Recommended Reading section) offer remarkably valuable perspectives on these issues. But, what I wanted to offer Megan was a recommendation that put all of the extraordinary challenges that her generation will face into some meaningful context, and balanced the various ideological viewpoints in a way that allowed her to reach independent conclusions. In sum, I wanted to provide her with a short-cut to understanding a debate that might otherwise take years to sort out.

Because that book had not yet been written, I sat down the next morning, began what was to become a self-imposed (and at times painful) one-year sabbatical from the NFL, golf, and YouTube, and wrote the pages that follow. It is my sincere hope that some day, when a parent is asked some version of the question Megan asked me on that crisp November afternoon, this book will be helpful and meaningful.

* * *

On November 20, 2009, an unidentified hacker stole thousands of emails from the University of East Anglia's Climate Research Unit (CRU) in eastern England. Because the CRU is among the world's most prominent institutes for the study of global warming, the publication of these private emails provided a rare glimpse into the inner sanctum of those scientists who champion what is commonly called the "consensus view" on climate change. The overwhelming majority of the hacked documents were of the relatively mundane, nose-to-the-scientific-grindstone variety. But a handful of these emails seemed to confirm the worst suspicions of global warming "skeptics" outside of the scientific mainstream.

In the emails, one of the world's most powerful climate scientists appears to encourage the deletion of documents to frustrate Freedom of Information Act (FOIA) requests, and others seem to be manipulating data to yield preferred results. Even worse, the emails demonstrate repeated attempts to suppress opposing viewpoints on the causes and dangers of global warming. In one such email, the CRU's Director, Phil Jones, writes to Michael Mann (a prominent climatologist at Penn State University): "I can't see either of these [skeptical] papers being in the next IPCC [United Nations Climate] report. Kevin and I will keep them out somehow even if we have to redefine what the peer-review literature is!"[1]

Public reaction to the emails fell largely along ideological lines. A number of skeptics' websites characterized the emails as a "smoking gun" that proved that climate change is a hoax or the "final nail in the global warming coffin."[2] At the other end of the spectrum, U.S. Climate "Czar" Carol Browner merely shrugged off the emails, indicating that she considered the science of global warming already well settled.[3] Not all reaction was so polarized, however, and certain voices of reason emerged, from both sides of the debate, concerning what might be learned from the emails. George Monbiot, a leading advocate of the consensus view, was among the first environmentalists to openly acknowledge their significance:

> I have seldom felt so alone. Confronted with crisis, most of the environmentalists I know have gone into denial. The emails hacked from the Climactic Research Unit (CRU) at the University of East Anglia, they say, are a storm in a tea cup, no big deal, exaggerated out of all recognition. It is true that climate change deniers have made claims which the material can't possibly support (the end of global warming, the death of climate science). But it is also true that the emails are very damaging. ... No one has been as badly let down by the revelations in these emails as those of us who have championed the science.[4]

From the other side of the climate aisle, an equally candid voice of reason was posted on the skeptics' climate blog: *The Air Vent*:

> As a skeptic, I can say in no uncertain terms that the emails and documents from the University of East Anglia do not show that [man-made global warming] is a falsehood or hoax. Claims that "global warming is dead" (as I have seen) are not supported by those documents. On the other hand, claims that "the science is settled" are shown to be an exaggeration.[5]

The disclosure of the East Anglia emails, and the divided reaction to them, represent a microcosm of the highly-charged global warming debate. Yet, the existence of reasonable "concessions" by partisan commentators demonstrate that even amid the deep cynicism and polarization that pervade the climate change controversy, there are scientists, journalists and observers willing to judge each issue on its own merits. And, just as some interested parties were able to put the meaning of the East Anglia emails into a reasonable context, there is common ground to be found on many of the other issues that form the larger global warming debate.

Introduction

Since the first single-celled organisms bathed in primordial soup 3.8 billion years ago, the relationship between climate and life has been one of active and reactive partners. But now, according to the world's leading scientific institutions, that prehistoric relationship is being redefined. The activities of human beings, and in particular man's combustion of carbon-intense fossil fuels, are thought to be warming the Earth, expanding the seas, and changing global precipitation patterns. What's more, many scientists warn that without drastic action, the Earth's ecosystems could soon be pushed past an irreversible tipping point.

By no means, however, is this "consensus" view the only view. Many highly regarded scientists and powerful policymakers believe that the jury is still out on whether human beings are causing the planet to warm. These individuals point out that the Earth's climate is always changing, that our current understanding of the climate system remains primitive, and that so-called political correctness has slammed the door on robust debate.

Beyond the enormously complex threshold question of whether anthropogenic (man-made) climate change is occurring, lies the equally complicated issue of what might be done to slow its effects. Each of the world's 193 nations would be impacted differently by a warmer world; each bears different historic responsibility for carbon emissions; and each has different access to sustainable

1

energy options. As cheerily summed up by one climate commentator, global warming is "an international, intergenerational collective action problem to be faced in conditions of uncertainty, distrust and short-termism."[6]

To complicate matters further, the possibility of dangerous, life-altering climate change has arrived just as two other sustainability dilemmas—human population expansion and dwindling global oil reserves—seem to be settling in. Challenges such as these—that must be faced with incomplete, contradictory and ever-changing information; without a central governing authority to provide order; and with time running out—have been given the name *wicked problems*. As we will see in the pages that follow, this designation particularly befits the quandary of global warming.

There are dozens of books and thousands of articles that effectively examine aspects of climate change, sustainability, or alternative-energy solutions. There are few, however, that seek to put these interrelated challenges into a meaningful context, that is, to "show the forest for the trees." There is value in understanding phenomena such as the greenhouse effect, peak oil, wind power, El Niño, and the Bali Roadmap, but to comprehend how they interact with each other is to see our world in a whole new light.

Besides exploring how the many pieces of the climate puzzle fit together, this book seeks to distinguish itself in one other critical respect: Its goal is to inform rather than to persuade. It has not been written to reinforce liberal, conservative, green, or skeptical views. No viewpoint has been ignored, and no opinion has been belittled simply because it is not widely accepted. In short, this book is intended to step back from the fray and to provide a fresh and balanced assessment of a debate that is all too often dominated by the extremes.

1

Three Keys to Unlocking the Mysteries of Climate Change

Humanity stands at a crossroads. According to National Academies of Science throughout the world, we are altering the climate. Each time we decide to cool a building, use an electrical appliance, or drive a car, heat-trapping carbon dioxide (CO_2) is added to the air. Atmospheric CO_2 now exceeds the level found at any point in the last 800,000 years,[7] and average global temperatures are rising at an accelerated rate. If these circumstances were not dire enough, world population growth; resource depletion; the increasing energy appetites of developing nations; and the specter of warming "feedback loops" threaten to lead to disastrous, irreversible consequences.

However, the story of global warming is not all gloom and doom. We come upon this great test of human ingenuity and fortitude just as we are developing the wherewithal to effectively respond. The chapters that follow will examine our species' great race against calamity. In Chapters 2 through 7, we explore the complex and evolving challenges of climate change (and consider the countervailing view that there is no "problem"). Chapters 8 through 14 outline and evaluate climate-change solutions, including "clean coal," nuclear power, electric cars, smart grid technologies, and promising renewable energies such as wind, solar, biofuel, and geothermal. Before setting off toward the perilous and unsettled future that awaits us, however, we begin with three fundamental tenets that will help guide us during our journey.

BIAS AND PROPAGANDA INUNDATE BOTH "SIDES" OF THE CLIMATE CHANGE DEBATE

It is now all but inevitable that government response to the threat of climate change will result in the largest redistribution of wealth and technology in history. As was the case with the changes spurred by World War II and the Information Age, the world's transition to renewable energy sources will produce spectacular winners and colossal losers. Hanging in the balance are the conflicting fates of established fossil fuel conglomerates and newly minted, well-capitalized, renewable energy companies. Indeed, a list of the business sectors that would be most positively or negatively impacted by government-imposed climate change solutions reads like a Who's Who of the most powerful industries in the world: finance, energy, transportation, manufacturing, agriculture and insurance.

Potentially affected corporations have used all of the tools at their disposal—lobbyists, advertising, the "media training" of sympathetic scientists, contributions to supportive think-tanks, and lots of political donations—to influence public opinion on the issues of climate change and sustainability.* Although individual corporations often take a more moderate tone, with billions of dollars at stake, their surrogates wage a public relations war in the traditional media and on the Internet.

In addition to pitting some of the titans of industry against each other, proposed climate change solutions contain a divisive ideological component. If climate change is acknowledged as a global problem, it follows logically that it will require a global solution. In turn, it follows that a global solution will require an international carbon bureaucracy that would inevitably infringe upon individual rights and national sovereignty. Many people who do not view global governance and international wealth distribution as worthy goals have grown concerned that the threat of climate

---

* In 2008 alone, more than a quarter of a billion dollars was spent on lobbying expenditures by the electric utility industry ($159.7 million) and the oil and gas industry ($130 million). Anne Mulkern, "Energy Companies Opened Wallets Wide To Sway House Climate Bill," The *New York Times*, July 23, 2009.

change has been co-opted as a means to engineer those larger political objectives.

The fear that some would use global warming as a tool to advance social agendas and to consolidate power is not without some justification. Many of the strongest advocates of an international emissions program would be political and/or economic beneficiaries of the expansive government infrastructure that such a program would require. Indeed, some proponents of more centralized global governance openly recognize that international climate agreements like the Kyoto Protocol have the capacity to serve as a gateway to that broader purpose. For example, on November 20, 2000, in an address to the United Nations Framework Convention on Climate Change, French president Jacques Chirac expressed the following views regarding the Kyoto Protocol's role as the first step toward a more powerful transnational institution: "For the first time, humanity is instituting a genuine instrument of global governance, one that should find a place within the World Environmental Organization . . . [this is] the first component of an authentic global governance."

As with climate solution advocates, many climate change skeptics* are motivated by self-interest. In an infamous memo dated April 3, 1998, oil industry representatives laid out a Draft Global Climate Science Communication Action Plan that would:

- set a project goal of convincing the American public that significant uncertainties exist in climate science;

- concluded that "victory will be achieved" when "average citizens 'understand' (recognize) uncertainties in climate science; recognition of uncertainties becomes part of the 'conventional wisdom' ";

- called for the recruitment of scientists for a media outreach program to establish the desired "uncertainty"; and

---

* Denigration of an opposing viewpoint by use of a derogatory label is an ancient propaganda technique. The climate change debate is awash in such epithets. If you express skepticism, you may be labeled a "flat earther," a "denier," an "anti-environmentalist" or part of a fringe element. If you express a preference for action, you may be an "alarmist" or the victim of a hoax. In an effort to afford respect to all involved, we will refer to these respective viewpoints as "climate change skeptics" and "climate solution advocates."

- called for the establishment of a Global Climate Science Data Center to provide logistical and moral support to those scientists with views differing from the those of the United Nations' climate scientists.

By 2002, consistent with this plan, oil industry representatives were investing millions of dollars into various policy think tanks and institutes that supported their preferred, skeptical message— including the American Enterprise Institute (AEI).[8] In February 2007, it was disclosed that the AEI had sent letters to scientists offering $10,000 to highlight the shortcomings of a United Nation's climate change report.[9] The Union of Concerned Scientists has suggested that between 1998 and 2005, ExxonMobil alone paid $16 million to forty-three organizations involved in the global warming debate.[10]

None of this is meant to suggest that the scientists endorsed by Jacques Chirac or ExxonMobil are corrupt, or even wrong. The point is simply that various proposed climate change solutions, with their very real economic consequences, have ignited a new-age propaganda war. While there are certainly sources for unfiltered information on climate, renewable energy, and fossil fuels, there are also organizations with open—and sometimes disguised —agendas, circulating half-truths, hyperbole, disinformation, and stereotypes.

By way of example, in January 2009, someone turning on her computer might have been greeted by the following, seemingly incongruous, headlines:

CLIMATE CHANGE QUESTIONED AFTER 2008 TIPPED TO BE COOLEST YEAR OF [21ST] CENTURY[11]

NOAA: 2008 GLOBAL TEMPERATURE TIES AS EIGHTH WARMEST ON RECORD[12]

In fact, both statements are correct, but neither is entirely complete. Only by developing a broad understanding of the global warming phenomenon, and by accounting for the possible bias of a given information source, can we effectively recognize these determined attempts to manipulate us with incomplete truths.

MUCH ABOUT CLIMATE CHANGE REMAINS A MYSTERY

The basic mechanism of global warming—increased concentrations of certain gases cause higher temperatures by permitting light energy to enter our atmosphere while trapping heat energy—is well understood and almost universally accepted. What remains shrouded in uncertainty are critical details of how the Earth's climate systems will react to higher concentrations of industrial emissions. With the exception of work performed by a handful of pioneers who were ahead of their time (see Chapter 2), the science of climate change is still in its adolescence.

It wasn't until the 1970s that the possibility of anthropogenic (man-made) global warming first entered the consciousness of the mainstream scientific community.[13] In the last three decades, the number of scientists studying climate change has grown exponentially. Climatologists, geologists, biologists, astronomers, physicists, and meteorologists have all contributed to a surge of new discoveries and the development of extraordinarily sophisticated climate models. Yet, even with these rapid advances, much about the Earth's complex and inter-dependent climate systems remains unknown.

We don't know the degree to which man-made emissions will trap heat. We don't know how climate feedbacks (such as changes in cloud cover) might accelerate or offset global warming. We don't fully understand the effect of ocean oscillations, sun flares, or changes in the Earth's elliptical orbit. We don't know if or when the Earth's great carbon sinks (discussed in Chapter 2) might overflow. And we don't know how temperature changes will affect other climactic phenomena such as hurricanes and rising sea levels. As Dr. David Schindler, an ecologist and climate change scientist from the University of Alberta, summed up:

> To a patient scientist, the unfolding greenhouse mystery is far more exciting than the plot of the best mystery novel. But it is slow reading, with new clues sometimes not appearing for several years. Impatience increases when one realizes that it is

not the fate of some fictional character, but of our planet and species, which hangs in the balance as the great carbon mystery unfolds at a seemingly glacial pace.[14]

So why, if so much remains undiscovered, are we inundated with "scientific conclusions" on how much climate change has occurred, how that change correlates to man-made emissions, and what the future manifestations of our altered atmosphere will be? The answer is simple: opinions about what *could be* occurring are generally less likely to make headlines or modify behavior than are declarations about what *is* happening. As a consequence, some issues are reported as "settled" long before they actually are. Exaggerated or unconfirmed results are sometimes condoned as a means to a desired end. As climate solution advocate Al Gore acknowledged in a 2006 interview, many believe some manipulation of the general population is justified:

> Nobody is interested in solutions if they don't think there's a problem. Given that starting point, I believe it is appropriate to have an over-representation of factual presentations on how dangerous it is, as a predicate for opening up the audience to listen to what the solutions are, and how hopeful it is that we are going to solve this crisis. Over time, that mix will change. As the country comes to more accept the reality of the crisis, there's going to be much more receptivity to a full-blown discussion of the solutions.[15]

Stanford University climate scientist Stephen Schneider further explained the temptation of many to censure doubt and uncertainty: "[W]e need to get some broad based support, to capture the public's imagination. That, of course, means getting loads of media coverage. So we have to offer up scary scenarios, make simplified dramatic statements, and make little mention of any doubts we might have."[16]

The unfortunate fact is that much of the climate change puzzle remains unsolved. Even the most advanced climate models lack the sophistication to account for the myriad combinations of feedbacks and counterbalances that nature has in store for us. That is not to say, however, that we don't understand the basic mechanisms of the greenhouse effect, or that we don't know enough to be very concerned. The point is, those serious about understanding global warming must cast aside the common rhetoric of "we now know all we need to know" and embrace the notion that human understanding of our climate continues to evolve. To concede that we do not know *everything* is not to concede that we are powerless to act.

GLOBAL WARMING IS A "RISK MANAGEMENT" ISSUE WITH A SCIENTIFIC COMPONENT

The third key to understanding the overall dynamic of the climate change debate is to recognize that global warming is not merely a scientific issue. While any proposed solution will start with the science, it will inevitably present highly contentious political, economic, and social issues. How much are the world's citizens willing to sacrifice in terms of economic productivity to stabilize atmospheric greenhouse gas (GHG) levels? Should national emission restrictions be based upon a historic baseline (which would slow the modernization of China and India) or calculated on a per capita basis (which would disparately impede the economies of countries like the United States, Australia, and Canada)? Should established industrial powers pay compensation to developing nations in the form of "climate change adaptation costs" because their historic emissions have contributed to global warming? While each of these questions has a scientific component, none can be answered by science alone.

In their own way, the political and economic challenges of climate change are every bit as complex as the science. In a number of international agreements, the world's nations have acknowledged their "common but differentiated" responsibilities in combating global warming. These agreements reflect a general acknowledgment that wealthy nations will lead the way, both in terms of reducing emissions and in terms of technology transfer to developing nations. For all practical purposes, this means that for an international climate change treaty to get done, developed countries such as the United States will have to voluntarily accept the "triple whammy" of higher relative production costs, transfer of intellectual property, and loss of capital to foreign markets.

Such a transfer of wealth to foreign competitors presents formidable political and economic challenges, which explains why the Kyoto Protocol (detailed in Chapter 7) was soundly rejected by the United States Senate.* This political impediment to an international solution was aptly summed up by columnist George Monbiot: "When you warn people about the dangers of climate change, they call you a saint. When you explain what needs to be done to stop it, they call you a communist."

Reduced to its basic essence then, global warming is a risk management issue. The first step in solving any risk management problem is to identify and assess the level of peril associated with a "business as usual" scenario. Such an assessment requires scientific analysis, but also includes an evaluation of national security and economic risk components. The second step in addressing a risk management problem is comparing the overall cost of inaction to varying levels of action. Because of the uncertainty of the science and the difficulty of comparing today's economic sacrifice to tomorrow's human suffering, this analysis is more complicated than setting the deductible on your auto insurance policy—but the basic concept is the same. A third step, which is not universal to all risk management decisions, but necessary to any global warming solution, is determining who should pay any prudent risk mitigation cost. As discussed in Chapter 8, such a risk management analysis, inclusive of the scientific, economic, and political components of climate change, is currently taking place on the world stage under the auspices of the United Nations' Framework Convention on Climate Change (UNFCCC).

---

* Although the Kyoto Protocol was never formally voted upon by the Senate, the Byrd-Hagel Resolution, which passed 95–0 , made clear that the treaty would not be confirmed.

2

The Basics of Climate Change

Armed with our understanding that: (1) much of what we see and read is colored by subtle and not-so-subtle bias; (2) although climate science has progressed substantially in recent years, many uncertainties remain; and (3) any global warming solution will involve a complex confluence of economic, political, and scientific considerations, we are ready to roll up our sleeves and start sorting through the issues.

THE GREENHOUSE EFFECT

Most available energy is, in some sense, solar. Plants (biofuels) depend upon the sun for light and warmth. The flow of the Earth's atmosphere (wind energy) and the flow of water (hydro, tidal, and wave energy) are powered by the sun. Coal, gas, and oil (fossil fuel energy) consist of prehistoric plants and animals, representing deposits of stored solar energy. The wonder of our planet, however, is not the amount of solar heat or light bestowed upon it, but rather how the Earth has adapted to use the sun's energy. There are innumerable barren planets that have been showered with varying amounts of heat and light. The reason life is able to thrive on Earth is something called the "greenhouse effect."

The key to understanding the greenhouse effect is to recognize that the sun's energy enters the Earth's atmosphere in one form (light) and often leaves the atmosphere in another form (heat). When sunlight falls upon an object, some of the sunlight's energy is absorbed by the object, converted to heat energy, and then released. Like a greenhouse, or like a car with its windows shut on a sunny day, our atmosphere lets the light energy in and then retains some of the heat energy. In a greenhouse, it is the glass panes that allow the passage of light and retain the converted heat. In our atmosphere, it is the GHGs (water vapor, carbon dioxide, methane, nitrous oxide, and ozone) that trap and retain the warmth.

While much of the science of climate change is subject to vigorous debate, neither the existence, nor the importance, of the greenhouse effect is debated. It is fundamentally understood that without greenhouse gases in our atmosphere, the average temperature on Earth would be 60 degrees Fahrenheit cooler than it is now. Thus, absent the greenhouse effect, the mean global temperature would be $-4°F$ ($-20°C$)—the equivalent of an average January day in Anchorage, Alaska.

A look to Earth's neighboring planets provides some idea of what conditions would be like if our atmosphere had a different concentration of greenhouse gases. Mars, with its thin atmosphere, is a frozen wasteland at $-63°F$ ($-53°C$). Venus, with its thick blanket of atmospheric gases, is a boiling cauldron at $842°F$ ($450°C$). While some of this temperature differential can be explained by the planets' respective distances from the sun, much is attributable to their varying greenhouse gas concentrations.

OVERHEATING OUR GREENHOUSE

While GHGs are essential for life on Earth, it turns out that there can be too much of a good thing. Human beings are extracting billions of tons of GHGs, which have been harmlessly stored below the surface of the planet for 300 million years, and emitting them into the air. At the same time, we are clearing forests—particularly rain forests—that would otherwise serve as carbon sinks, absorbing massive amounts of carbon dioxide. Once released, these GHGs accumulate at the top of the troposphere (the lowest level of our atmosphere), where some will stay for hundreds of years. The core question then, is what effect, if any, will this increase in atmospheric GHGs have on the Earth's climate? The best place to begin our analysis of this critical issue is in Sweden, circa 1896.

A Swedish chemist named Svante Arrhenius (1859–1927) is credited for being the first scientist to propose that fossil fuel combustion would lead to increased carbon dioxide concentrations, which in turn would increase the Earth's average surface temperature. Arrhenius concluded that a doubling of atmospheric carbon dioxide could result in a 9°F (5°C) rise in temperature. Remarkably, this figure is in line with results produced by some of today's most complex climate models.

Unfortunately, Arrhenius's groundbreaking work on the issue gained little traction during his lifetime and the topic was all but forgotten until an American scientist named Charles Keeling began his own work six decades later. Keeling set out to determine if the first of Arrhenius's conclusions—that fossil fuel combustion would lead to increased atmospheric carbon dioxide—could be scientifically verified. In 1958, a fortuitous combination of available funding, technological advance, and a new summit road built by the army, permitted Keeling to take air samples from atop the Mauna Loa volcano on the Big Island of Hawaii.

Rising 13,679 feet out of the middle of the Pacific, Mauna Loa offered Keeling access to pristine air in the middle of the troposphere. For decades, Keeling withstood an array of financial and political challenges in order to continue his measurements. The result of his fortitude, the set of data now widely known as Keeling's Curve (see Figure 1 on page 14), marked an extraordinary breakthrough. In addition to scientifically establishing that atmospheric carbon dioxide was trending sharply upward, Keeling discovered something we now call "seasonal oscillation cycles." As demonstrated by the small, squiggly lines trending ever higher, each year carbon dioxide levels oscillate, increasing during the Northern Hemisphere's winter (when the trees are dormant), and decreasing in the summer (when the trees act as "carbon sinks").*

---

* The Northern Hemisphere dominates this carbon sink process because it holds much more landmass than the Southern Hemisphere.

The discovery of this annual phenomenon, sometimes described as "the breathing of the Earth," provides an important insight into the complex interconnectivity between the planet's systems.

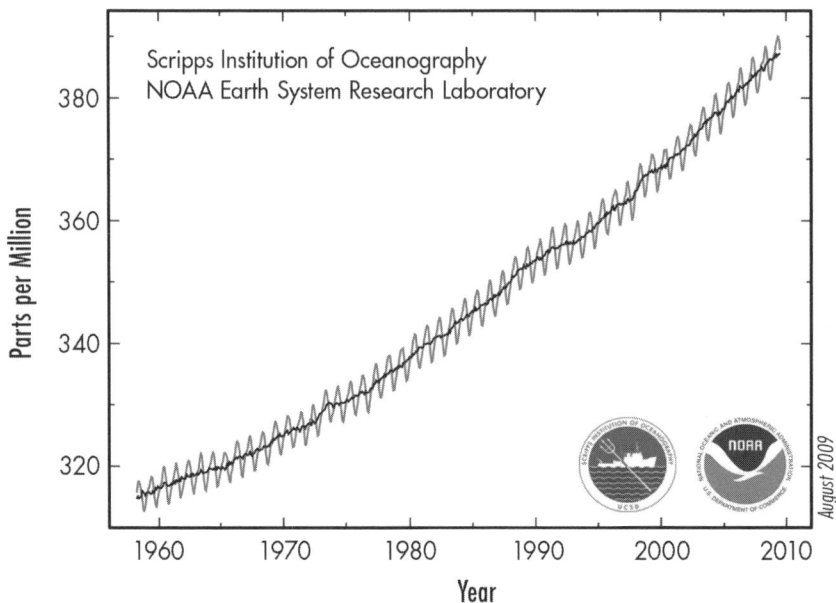

FIGURE 1. Keeling's Curve. Atmospheric CO_2 at Mauna Loa Observatory (courtesy of NOAA)

Keeling's discovery led a generation of scientists—chemists, meteorologists, physicists, biologists, oceanographers, and geologists—to revisit Arrhenius's theory of anthropogenic global warming. While this eclectic group of individuals has no singular voice, certain men and women have emerged as leaders, and none more so than James Hansen. A physicist by training and the director of NASA's Goddard Institute for Space Studies (GISS), Hansen became the first top climate scientist to publicly assert that global temperatures had risen beyond the range of natural variability. In his now famous testimony to Congress, which took place on a humid June afternoon in 1988, Hansen announced:

> I would like to draw three main conclusions. Number one, the Earth is warmer in 1988 than at any time in the history of instrumental measurements. Number two, the global warming is now large enough that we can ascribe with a high degree of confidence a cause and effect relationship to the greenhouse effect. And number three, our computer climate simulations indicate that the greenhouse effect is already large enough to begin to effect the probability of extreme events such as summer heat waves.[17]

Since his 1988 testimony, Hansen has been a lighting rod for both praise and criticism. He has received countless awards, and in 2006 was among *Time Magazine's 100 Most Influential People*. But in 2007, when a flaw was discovered in a GISS computer model (that resulted in an overvalue of surface temperature), Hansen was roundly excoriated by climate change skeptics as an irresponsible alarmist. In response to this 2007 "brouhaha," as he characterized it, Hansen offered the following judgment of his critics:

> [I]f tipping points are passed, if we, in effect, destroy Creation, passing on to our children, grandchildren, and the unborn a situation out of their control, the contrarians who work to deny and confuse will not be the principal culprits. The contrarians will be remembered as court jesters. . . . The real deal is this: the "royalty" controlling the court, the ones with the power, the ones with the ability to make a difference, with the ability to change our course, the ones who will live in infamy if we pass the tipping points, are the captains of industry, CEOs in fossil fuel companies such as EXXONMobil, automobile manufacturers, utilities, all of the leaders who have placed short-term profit above the fate of the planet and the well-being of our children.[18]

To this day, Hansen remains an important voice in the global warming debate, and his continuing interaction with climate change skeptics is often thought-provoking.

WHAT SORT OF CHANGES HAVE ALREADY OCCURRED?

Although we know that greenhouse gases have the capacity to warm our planet, and that atmospheric concentrations of carbon dioxide have grown by 40% over the last 250 years, great uncertainty still exists concerning the extent and manner in which these increases will impact the climate. The consensus view is expressed in terms of likely future temperature ranges and is based upon complex climate models which, by necessity, include some inferential component. Weighing interactive factors such as volcanic activity, solar variation, ocean oscillation (such as El Niño), and the competing effects of greater cloud cover (increased sun reflection) and reduced ice cover (decreased sun reflection) in a warmer world, all require some level of speculation. Given the lack of an available "test planet" with which to experiment, the measurable changes on Earth are critical, not just in their own right, but also as a means to test the predictive capacity of the various models. So what, if any, perceptible changes have we actually seen to date? Three examples—temperature increases, melting ice, and rising sea levels—are discussed here.

In the last century, the global mean surface temperature has increased 1°F (0.6°C), with most of that change occurring in the last three decades (Figure 2). Since the mid-1970s, the warming has quickened to a disquieting rate of .32°F per decade (3.2°F per century). The eight warmest years since 1850 (when methodical thermometer records first became available) have all occurred since 1998, with 2005 being the warmest year.[19]

These quantifiable temperature increases are consistent with the theory of anthropogenic global warming, and *suggest* a human "forcing" of the climate, but do they *prove* that the forcing was not caused by natural factors? That is, quite literally, the trillion-dollar question. According to the Intergovernmental Panel on Climate Change (which is regarded by the United Nations and many in the media as an authoritative voice on climate science), the answer is "probably." Specifically, a 2007 IPCC summary report concluded:

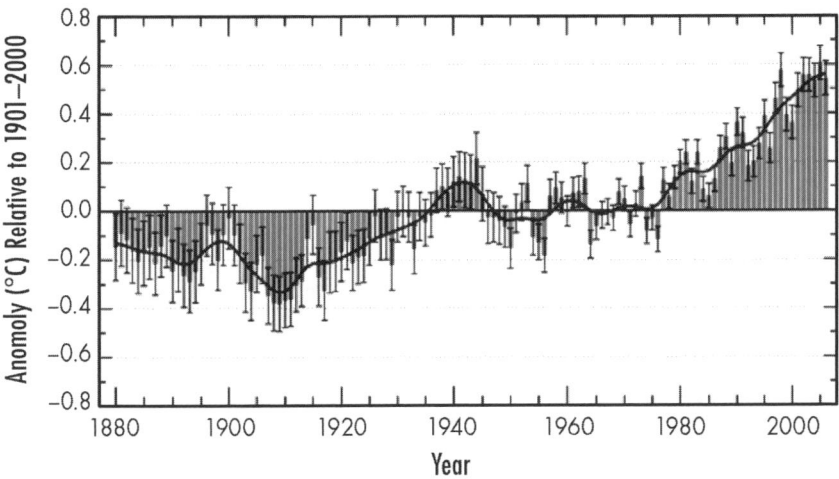

FIGURE 2. Jan–Dec Global Mean Temperature Over Land and Ocean (*courtesy of NOAA*)

> The observed widespread warming of the atmosphere and ocean, together with ice mass loss, support the conclusion that it is *extremely unlikely* [less than a 5% chance] that global climate change of the past 50 years can be explained without external forcing, and *very likely* [greater than a 90% chance] that it is not due to known natural causes alone.[20]

As discussed in Chapter 6, climate change skeptics: (1) take issue with the reliability of the models that the IPCC utilizes to make these assessments; and (2) point out that these seemingly definitive conclusions explicitly carve out both the possibility of a non-anthropogenic external forcing (such as solar variations), and the possibility of a natural forcing from a cause not yet known to the fledgling science of climate change.

Recent temperature increases have had a demonstrable and alarming effect on melting sea ice. Arctic temperatures have increased at a rate almost double that of the global average over the last century. In recent years, that warming has substantially reduced both the ice extent (total area covered by some ice) and the ice thickness in the Arctic. Each year, sometime in mid-September, the Arctic Sea ice extent reaches its annual low. Since satellite records became available in 1979, the summer extent has progressively trended downward (Figure 3), leading scientists to conclude that the Arctic Sea might become ice-free by the mid-twenty-first century. Then, in 2007, something extraordinary occurred. On September 16, 2007, the ice extent fell to 4.13 million square kilometers—25 percent less than any other measured year. This abrupt decline has led to numerous predictions of a seasonal ice-free Arctic between 2012 and 2020.[21]

Although the relatively cooler 2008 temperatures resulted in a slight "recovery" of surface ice area, the 2008 extent, at around 4.5 million square kilometers, was the second-lowest extent since such measurements became available in 1979.* Developing data also suggests that the total *volume* of ice was actually less in 2008 than in 2007, because the 2008 ice was *thinner*.[22] Because the depth of the ice reveals its susceptibility to melt, thinning can be viewed as a leading indicator of shrinking surface ice. Precise quantification of

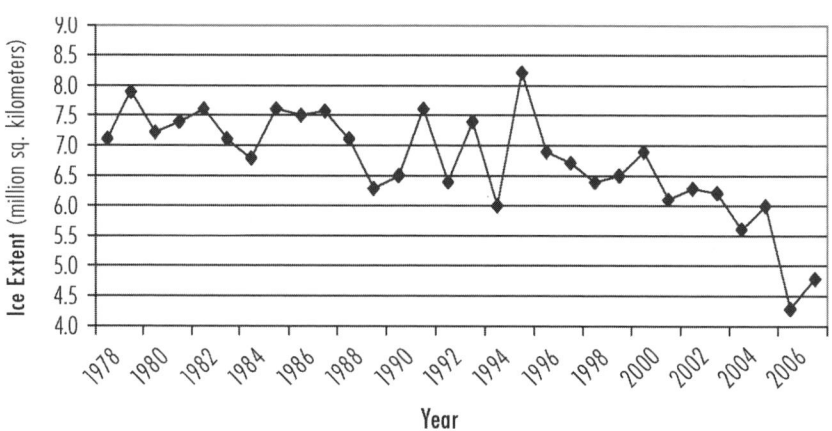

FIGURE 3. Decline in September Arctic Sea Ice *(courtesy of National Snow and Ice Data Center)*

---

*On September 12, 2009, the Artic extent reached 5.1 million square kilometers, making the 2009 extent the third lowest since 1979.

Arctic ice thickness remains elusive because of its variability throughout the region, and because satellites are much more effective at measuring surface ice. But with the help of special satellite technology and data collected by navy submarines, NASA's ICE-Sat program has estimated that, between 2004 and 2008, the thickness of winter sea ice decreased at a rate of 7 inches per year.[23]

In addition to melting sea ice, the Earth's land surface ice—in the form of glaciers and ice sheets—has also contracted. Because glaciers are many magnitudes smaller than ice sheets, and because they are dispersed throughout the world (on every continent except Australia), glaciers are seen as an important barometer of climate change. According to the World Glacier Monitoring Program, which has measured glaciers around the world since 1980, more glacier ice has been lost than gained in every year since 1989—with the rate of loss more than doubling since the 1990s.[24]

The Antarctic and Greenland Ice Sheets contain 99% of the planet's permanently frozen, terrestrial ice, and are much more resistant to change than glaciers. In Greenland, the area affected by ice melt on at least one day per year increased by 42% from 1979 to 2005, and in 2007 the melt extent exceeded the existing record by 10%. Data concerning the Antarctic Ice Sheet is less clear. While the spectacular collapse of the Rhode Island–sized Larsen B Ice Shelf in 2002 is often attributed, in part, to warmer temperatures, there is also evidence that the Antarctic's winter ice is expanding.[25] In 2007, the IPCC concluded that too little was known about Antarctica to draw definitive conclusions on the effects of global warming upon that continent. Yet, by 2009, researchers at NASA and the University of Washington, through the use of new weather station data and satellite records, had concluded that the Antarctic continent as a whole had warmed.[26]

Another measurable impact of the Earth's warming climate is its rising seas. Ocean levels have risen about 8 inches (20 cm) in the last 130 years as a result of increased global temperatures, and the expansion of the seas is accelerating. Both the causes and the potentially devastating effects of this trend are discussed in Chapter 4.

CARBON SINKS

One of the more intriguing aspects of the Earth's climate system is the operation of its great carbon sinks. Only 40% of the CO_2 emitted into the air reaches the top of the troposphere and contributes to the greenhouse effect. The remainder is absorbed, in almost equal parts, by oceans and plants—substantially slowing the effects of global warming. Precisely how these sinks work, and whether they are reaching overflow capacity, are the subjects of vigorous study and debate.

The world's oceans absorb approximately 30% of the CO_2 released when humans burn fossil fuels. Almost half of that absorption occurs in the Southern Ocean,* near Antarctica, because carbon dioxide dissolves more efficiently in cooler waters. Two mechanisms drive the absorption process. The first is the ocean's tendency to absorb more carbon as the CO_2 levels in the air increase. Generally speaking, an increase of a given substance in the atmosphere will mean more of that substance is dissolved into the ocean. The second absorption mechanism is biological. Phytoplanktons (algae) in the ocean use photosynthesis to extract carbon from CO_2. The bloomed plankton is consumed by marine life and ultimately falls to the seabed in the form of skeletons and shells.

The Southern Ocean is full of all of the nutrients necessary for phytoplanktons to flourish except one: iron. As a consequence, one of the more promising areas of climate research involves the possibility of inhibiting the greenhouse effect by fertilizing the ocean surrounding Antarctica with iron. A recent study has suggested that between ten thousand and one hundred thousand carbon atoms are absorbed for each iron atom added to the water.[27] Thus, large tankers fertilizing that area with iron sulfate could potentially remove substantial levels of carbon from the atmosphere.

---

* For those of you, like the author, who last attended science class some time in the twentieth century, the International Hydrographic Organization designated the southern portions of the Atlantic, Pacific, and Indian Oceans as the world's fifth ocean—the Southern Ocean—in 2000.

Despite great promise, however, the risk that such geoengineering would upend the ecosystem led the United Nations Convention on Biological Diversity to call for a temporary moratorium on any such plans.[28]

The second type of carbon sink goes by many names—the terrestrial eco-system, the land biosphere, or simply forest area—but whatever it is called, it operates primarily through the natural ability of trees (and to a lesser extent crops and grasslands) to store carbon. It has long been believed that higher levels of ambient carbon dioxide have a fertilization effect on plant life and that as much as 30% of our increased carbon emissions have been absorbed by trees and other flora. Reforestation in North America has been of particular benefit because young, growing forests are even greater net consumers of carbon than mature forests. Conversely, deforestation, particularly the clearing of carbon rich rain forests, releases vast amounts of CO_2 back into the atmosphere.

There is vigorous ongoing debate over whether one, or perhaps both, of these two great carbon sinks may be losing its ability to absorb carbon. In 2007, there were widespread reports that the Southern Ocean was losing its storage capacity—and that as a result, "carbon dioxide could accumulate in the atmosphere faster than expected over the coming decades."[29] In 2008, however, a study performed by Australian and German scientists, with access to additional data, reached a contrary conclusion, finding that the Southern Ocean's ability to soak up carbon dioxide has not been altered in recent years.[30] Currently, there is no definitive understanding of whether the capacity of the ocean sink has diminished or may diminish in the future.

The continuing ability of the terrestrial sink to absorb 30% of carbon emissions is also in doubt. It is generally assumed that increased atmospheric CO_2 will spur growth by promoting photosynthesis. There is a growing concern, however, that this fertilization effect will be offset (in whole or in part) by either deforestation or a slower growth rate of tropical trees attributable to a warmer, more arid climate.* Thus, while we know how both the ocean and terrestrial carbon sinks have operated in the past, the possible saturation of these sinks is one of climate science's great unknowns.

---

* The concern that the rainforests might not survive a substantial temperature increase appears to have been quelled by the discovery of a 60-million-year-old snake. The "titanaboa," which grew to be 43 feet long and dieted on crocodiles, lived in South America at a time when temperatures were 5°C warmer than the present, suggesting that the rainforest could indeed thrive in much warmer conditions.

3

Feedback Loops
Nature's Risk Amplifiers

The Earth's climate is a dynamic system, ever changing over time. The speed, extent, and nature of that change is profoundly influenced by a variety of competing positive and negative climate feedback mechanisms. A negative feedback is a stabilizing force tending to keep a biological, chemical, social, or other system in equilibrium. By way of example, the population of a predator species is generally controlled by a negative feedback mechanism. As the population of the predator expands, the availability of its prey is reduced, and starvation reduces the predator's numbers. Conversely, a positive feedback amplifies the original effect, driving change and destabilizing the system. The interaction between human population growth and technological advance is an example of a positive feedback loop, each output reinforcing the other. Understanding our weather system's many feedbacks, and then successfully integrating their competing effects, represents a formidable challenge facing today's climate scientists.

THE ICE-ALBEDO EFFECT

The measure of an object's propensity to reflect light energy is known as its *albedo*. A perfectly dark material has an albedo of 0%, which means it reflects no light and absorbs it completely. Conversely, an absolute white material has an albedo of 100% (complete reflection and no absorption). As a result, vegetation, soil, and water (with albedos between 10% and 20%) absorb much more heat than ice and snow do (with albedos of 90%). See Figure 4.

FIGURE 4. SAMPLE ALBEDOS			
Pure White Surface	1.0	Average of Earth's Surface	0.3
Fresh Snow	0.9	Grass Field	0.2
Sea Ice	0.8	Open Water	0.1–0.15
Desert Sand	0.4	Pure Black Surface	0.0

When global temperatures drop, a larger portion of the Earth's surface becomes covered with ice. In turn, the newly formed ice reflects sunlight back into space that had previously been absorbed by darker oceans and land. The reduction in solar absorption pushes temperatures even lower, causing a further expansion of reflective ice, and the positive ice-albedo feedback loop is off and running. The conventional scientific view is that this ice-albedo amplification effect is a crucial factor in the periodic occurrences of ice ages.*

Importantly for our purposes here, the ice-albedo effect also runs in reverse. Warmer temperatures melt ice and snow cover, exposing more land and sea. The new darker surfaces reflect less solar heat back into space, soaking up more warmth. The heated atmosphere melts more ice and snow, which causes more warming, and a self-reinforcing cycle has been set in motion.

THE PERMAFROST/CARBON FEEDBACK

Another positive feedback loop with runaway potential is permafrost/carbon amplification. Permafrost occupies about 24% of

---

* The book *Snowball Earth: The Story of the Great Global Catastrophe that Spawned Life as We Know It*, by Gabrielle Walker, provides an insightful discussion of the primary and secondary causes of ice ages—as well as possible explanations of how the ice-albedo effect is reversed at the end of an ice age.

the exposed land surface in North America. Although defined as any ground that has remained frozen for at least two years, most of the world's permafrost is thousands of years old. Trapped within this cryogenic layer are large deposits of carbon in the form of long dead plants and animals. When that organic material decomposes in the air, it forms carbon dioxide, and when it decomposes under water, it forms methane. Approximately 1500 billion tons of carbon are stored in the Earth's permafrost, roughly double the carbon that is now contained in the atmosphere.[31]

As permafrost ground and bogs thaw, the heat-trapping carbon dioxide and methane are released into the atmosphere. This increase in atmospheric greenhouse gases causes further warming, which melts more permafrost, releasing further GHGs and the makings of a potentially vicious loop have been put in motion (see Figure 5).

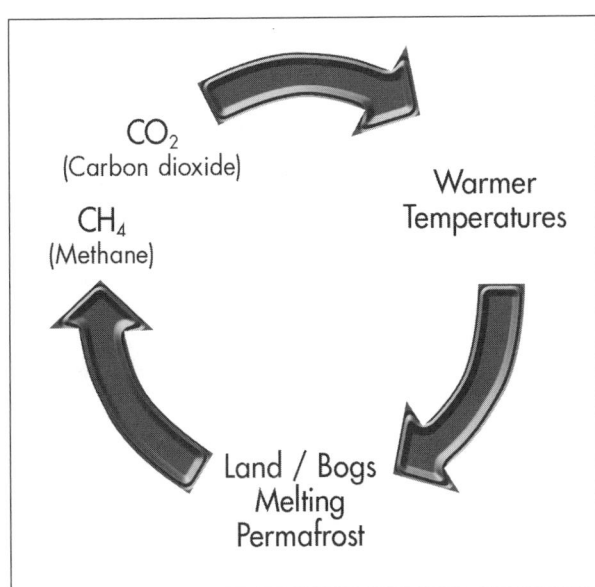

CO_2
(Carbon dioxide)

CH_4
(Methane)

Warmer
Temperatures

Land / Bogs
Melting
Permafrost

FIGURE 5.
The Permafrost/
Carbon Feedback
Loop

The permafrost/carbon feedback loop is of particular concern because of the potentially large amount of methane that could be released. Although methane's ten-year lifespan as a greenhouse gas is much shorter than that of carbon dioxide,* methane is twenty-five times more effective at trapping heat. As a result, the release of methane, particularly from Siberia's vast frozen peat bogs, is one of the climate tipping points most feared by scientists.[32]

WATER VAPOR, CLOUDS, AND CONFUSION

The role of water vapor as a climate feedback mechanism is a hotly contested subject. This issue, as much or more than any other, exemplifies how much we *don't* know about the workings of climate change. What scientists do know is that a warmer atmosphere pulls more water vapor from the ocean, and that increased atmospheric water vapor results in more humidity and cloud formation. Beyond that, there are many opinions and little certainty or consensus.

On one hand, water vapor is a powerful greenhouse gas so an increase in humidity will trap heat and amplify an increase in global temperature. A report issued by Texas A&M scientists in November 2008 concluded that this water vapor feedback alone is capable of doubling the warming caused by carbon dioxide.[33]

On the other hand, increased cloud formation caused by a higher concentration of atmospheric water will act as a cooling mechanism. Clouds account for more than half of the Earth's albedo effect, reflecting about one sixth of all incoming sunlight back into space.[34]

So which of these competing water vapor feedback effects will ultimately dominate the other? Most climate models include an assumption that increased water vapor will have a net amplifying effect on global warming. The view that water, on balance, is a positive feedback mechanism, however, is far from universal. As succinctly expressed by Working Group I of the UN's Intergovernmental Panel on Climate Change (IPCC) in 2007: "Cloud feedbacks remain the largest source of [climate system] uncertainty."[35]

---

* The atmospheric lifetimes of GHGs vary greatly: carbon dioxide (hundreds of years), methane (12 years), nitrous oxide (114 years), and hydrofluorocarbon-23 (264 years). Additional information on GHG lifetimes can be found in Table 2.14 of the IPCC AR4 WG-1 Report.

OTHER FEEDBACKS

Climate models must also account for a variety of other feedbacks that will either intensify or inhibit rising temperatures. Possible environmental responses to a warming planet include the following:

- **Forest Dieback:** As temperatures increase and precipitation patterns change, some forests will become stressed and more susceptible to fire and pests. The dieback of these ecosystems will release carbon into the atmosphere, putting in motion a positive feedback loop.

- **Plankton Absorption:** As discussed in Chapter 2, plankton absorbs large quantities of carbon before sinking harmlessly to the ocean floor. Some studies suggest that warmer ocean temperatures block the flow of nutrients required for plankton growth thus amplifying the warming effect. Conversely, increased temperatures and melting will release iron trapped in icebergs into the ocean causing plankton to bloom thus dampening the warming effect.

Climate feedback mechanisms are complex and integrated, and they continue to befuddle scientists. This uncertainty detracts from the reliability of climate models and serves as fodder for climate change skeptics. To some degree, these criticisms have merit. Water vapor is inputted into most models as a net positive feedback, but as the IPCC's 2007 assessment points out, the impact of increased cloud formation remains largely unknown. The uncertainty, however, cuts both ways. As discussed in Chapter 5, the Earth's paleoclimatic archives demonstrate that the possibility of runaway climate change, induced by positive feedback loops, should be taken very seriously.

4

El Niño, Volcanoes, and Global Dimming
The Climate Change Wildcards

If climate feedbacks are Mother Nature's curveball, it turns out that she also throws a pretty nasty slider, change-up, and screwball. This chapter addresses two prominent climatologic factors that are exerting their own pressure on the weather, and a third factor that may be temporarily masking the effects of global warming.

EL NIÑO AND VERY WARM WEATHER

The most severe climactic extremes of the last few decades have occurred during periods of El Niño. As a result, determining how El Niño events affect temperature and how temperature, in turn, affects these events is a critical piece of the climate change puzzle.

El Niño, and its counterpart La Niña, are part of an atmospheric and oceanic cycle known as the El Niño Southern Oscillation (ENSO). Under ordinary ENSO conditions, reliable trade winds along the Pacific equator push massive volumes of water from the South American coast toward Asia. The water that is blown from east to west across the Pacific Ocean remains near the surface where it is drenched with the warmth of the sun. When this warm water comes to rest along the coasts of Indonesia and northern Australia, it constitutes the world's largest heated pool. In stark contrast, along the coasts of Ecuador and Peru, the evacuated, wind-blown surface water is replaced with chilled water that has been cycled up from the ocean's depths. As a consequence of this ENSO effect, equatorial waters in the western Pacific (between Australia and Asia) are 14°F (8°C) warmer (and piled a half meter higher) than equatorial waters in the eastern Pacific. Under ordinary ENSO conditions, the cooler waters extend out from the South American coast in an oval shape known as the "cold tongue."

Once every three to seven years, the westward trade winds slow, and the warm pooled water pours eastward back across the Pacific Ocean toward the Americas. This phenomenon was dubbed "El Niño," or "the Christ child," by Peruvian fishermen who noticed that the warming often occurred around Christmas time. El Niño conditions rarely last more than one year and are often immediately followed by a La Niña event. During La Niña, the trade winds are particularly strong, more surface water is pushed westward, more cold water is pulled from the ocean depths along the South American coast, and the cold tongue expands.

During bouts of El Niño, when the warmer waters escape the pool and spill across a greater area of the ocean's surface, the low-level atmosphere is heated. The warmer air causes an increase in average global temperature, precipitation changes, and a variety of extreme weather events. While El Niño does not explain the long-term global warming trend, it does account for short-term temperature spikes. The 1997–1998 El Niño, one of the strongest ever recorded, was followed by record high global temperatures, an increase in extreme storms, and severe droughts. In La Niña years, when vast amounts of cool water are pulled from the depths of the ocean, average global temperatures drop.

The variability of ENSO is one of many reasons global warming should be analyzed, not by focusing on individual years, but in the context of larger climate trends. The 2008 La Niña conditions present a prime example of how accurate temperature data, taken out of context, can be used to support misleading conclusions. In early 2009, the blogosphere was rife with articles characterizing 2008 as an unusually cool year. Two of the articles pronouncing the demise of global warming read:

2008 THE COLDEST YEAR OF THE CENTURY: SO MUCH FOR GLOBAL WARMING[36]

GLOBAL WARMING TAKES ANOTHER HIT— 2008 COLDEST YEAR OF THE DECADE[37]

In point of fact, 2008 was the coldest year of the young twenty-first century, but fair disclosure would have required an acknowledgement that 2008 was also the eighth-warmest year in recorded history—warmer than all but two years in the twentieth century. Full disclosure would have also required a revelation that 2008 temperatures were reduced by a strong La Niña. In other words, a year in which global temperatures were substantially cooled by prevailing ENSO conditions turned out to be one of the ten hottest in the last century and a half. Thus, these headlines, and hundreds of others like them, are most graciously characterized as incomplete.

While scientists have high confidence concerning how the various stages of ENSO affect global temperature, they know relatively little about how global warming might, in turn, affect ENSO. It does appear that El Niño and La Niña events have increased in frequency and severity in recent years—and some scientists, while acknowledging the lack of conclusive proof, have theorized that climate change may be the cause of those increases.[38] What is generally agreed upon among scientists is that climate phenomena like ENSO and the greenhouse effect are highly interactive systems, and that a change in one of these systems is likely to have some impact on the others. For this reason, the study of the relationship between an increasingly volatile ENSO effect on the one hand, and global warming and its many manifestations on the other, will be receiving substantial attention in the coming years.

VOLCANIC ERUPTIONS

Another key to determining the extent to which increased GHG concentrations are forcing global warming is successfully integrating the cooling effect of volcanic activity into current climate models. Some level of minor volcanic activity around the world is relatively constant, but larger eruptions, which occur only several times a century, spew enough ash to block out vast amounts of incoming solar radiation. Essentially, sulfur dioxide is ejected and propelled through the troposphere and into the stratosphere where it converts into sulfate aerosols. These stratospheric aerosols reflect the sun's energy, and the global cooling caused by major eruptions typically lasts two to three years.

The American winter of 1783 to 1784 was the coldest ever recorded—an astonishing 8°F (4.5°C) below the 225-year average.[39] As Benjamin Franklin recognized at the time, this bitter cold followed an extremely severe volcanic eruption in Laki, Iceland, that killed 20% of the Icelandic population and more than half of its livestock. Blown ash covered England, and the haze over Europe resulted in an untold number of sulfur dioxide inhalation deaths. Franklin's hypothesis that the expelled atmospheric gases contributed to cooling in Europe and North America has since been confirmed by modern science.

In 1815, the Laki volcano was substantially outdone by the eruption at Mount Tambora in Indonesia.* Tambora, one of five known super-colossal volcanic events in the last ten thousand years, exploded with roughly one hundred times the force of the 1980 Mount Saint Helens eruption in Washington state. The explosion sheared four thousand feet off of Mount Tambora, and an immense ash cloud caused cooler temperatures around the world. The following year, 1816, is widely referred to as "the year without a summer." Europe experienced crop failure and extreme famine, and in New England, half a world away from Indonesia, evening frost persisted into July and August.

In more recent years, two major volcanic eruptions have emitted enough sulfur dioxide to have a measurable impact on global temperature—El Chicon (Mexico, 1982) and Pinatubo (Indonesia, 1991). Figure 6 demonstrates both, the increased stratospheric aerosol

---

* For a comprehensive analysis on why and where volcanoes occur, and why the Indonesian archipelago is particularly vulnerable, see Simon Winchester's *Krakatoa: The Day the World Exploded.*

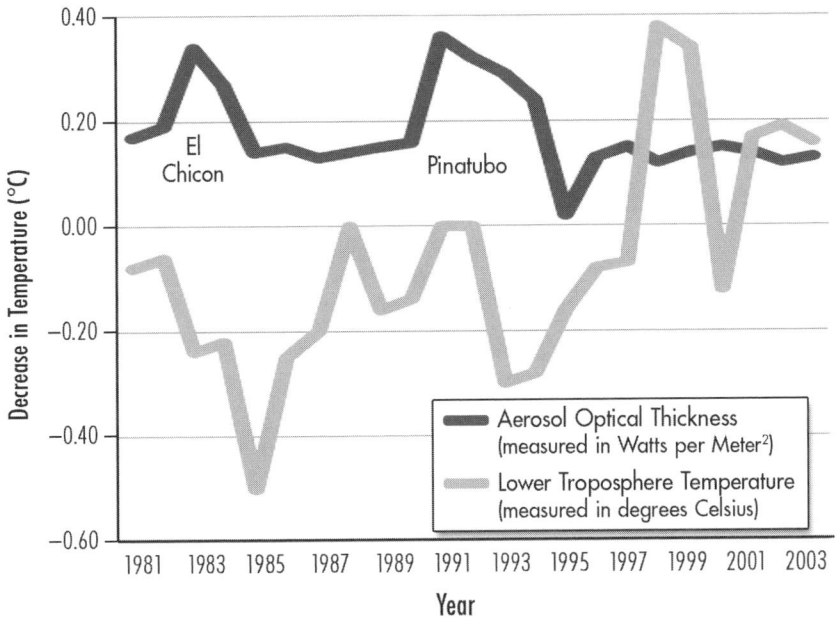

FIGURE 6. Decreased Atmospheric Temperatures Following Volcanic Activity (*Source: NOAA and NASA*)

levels caused by these eruptions, and the corresponding drop in global temperature in the two or three years following these events. What this graph does not show, however, is that the El Chichon and Pinatubo eruptions substantially lowered world temperatures during particularly strong El Niño years that would have otherwise resulted in exceptionally warm conditions.

GLOBAL DIMMING

At the turn of the twenty-first century, a scientist broaching the subject of global dimming with his peers would have elicited blank stares and raised eyebrows. With the publication of two ground-breaking articles in 2001 and 2002,[40] however, the radical notion—that the amount of sunlight reaching the Earth's surface has been steadily decreasing since 1950—has rocketed from the scientific fringes into the mainstream.

The theory of global dimming holds that the discharge of man-made aerosols into the atmosphere has caused a gradual reduction in the amount of direct sunlight that reaches the surface. Aerosols are nongaseous particles, usually solids, so small that they will stay aloft in the troposphere for hours, weeks, or even months. These particulates can enter the atmosphere through natural events, such as volcanic activity, wildfires, and sandstorms, or through the man-made discharge of pollution. Coal-fired power plants, automobiles, and industrial processes have added substantial levels of "unnatural" aerosols to the skies above us.*

Aerosol pollutants influence the climate in two ways. First, they absorb and reflect sunlight bound for the Earth's surface, causing a dimming in the lower atmosphere. This increase in the planet's overall albedo has a cooling effect. Second, aerosols can operate as a bonding agent, bringing water droplets together and forming clouds. As discussed in Chapter 3, clouds also have the capability to cool the planet by reflecting solar energy.

A majority of climate scientists now believe that global dimming's cooling effect has masked some of the warming effect of increased atmospheric GHGs. The extent that aerosols have offset global warming remains unclear. As one climate modeler from England's highly respected Hadley Centre explained in 2003, the incorporation of global dimming into the computer simulations remains "a bit of a work in progress."[41]

Despite uncertainty on many of the finer points, however, there is good reason to believe that aerosols released by man in the combustion of fossil fuels play a major role in the climate system. Measurements available from the 1950s demonstrate that, in each of the three decades between 1960 and 1990, sunlight reaching the Earth decreased by 4 to 5%—with much larger reductions in industrial regions of the world.[42] That dimming trend leveled off and slightly reversed around 1990—shortly after the United States and

---

* Unlike greenhouse gases that stay in the atmosphere long enough to be dispersed around the globe, aerosols often impact climate on a regional level. Nonetheless, plumes of aerosols, often in the form of atmospheric brown clouds (ABCs), can travel across oceans and continents.

European nations enacted clean air legislation substantially restricting aerosol emissions. Notably, China and India, which have not implemented such pollution controls, have seen "no upturn in [solar] radiation."[43]

So how much of a masking effect have aerosols and global dimming had on global warming? Although a soft estimate of as much as 50% is often floated in articles and on blogs,[44] this is yet another climate question that remains unanswered. There are, however, two intriguing clues which suggest that the cooling impact of aerosols is potentially substantial. First, in the 1990s, just as global dimming appeared to level off, world temperatures rose sharply. Second, there is evidence demonstrating that, when jet exhaust plumes are removed from an area of the sky, daytime temperatures in that region increase. Although the theory is not easily tested, an opportunity to do so arose out of the tragedy of the September 11, 2001, World Trade Center attacks. Following those events, civil air traffic over United States air space was banned for three consecutive days, and the jet contrails that would normally have streaked across the skies disappeared. Temperature measurements during those days revealed an abnormally high variance between warmer day and cooler night temperatures. This anomaly is viewed by many scientists as further confirmation that anthropogenic aerosols are a major factor in reducing daytime temperatures.

Having evaluated the theory of global warming and considered the measurable climatic changes that have already occurred in Chapter 2, and having acknowledged several important complicating factors in Chapters 3 and 4, we move to the future—where the disquieting realm of the probable and the possible awaits.

5

Future Manifestations of Global Warming
The Potential Costs of Inaction

"According to a new U.N. report, the global warming outlook is much worse than originally predicted. Which is pretty bad when they originally predicted it would destroy the planet."

—JAY LENO

Human beings have tailored their societies around existing ecosystems, sea levels, and coastlines. Alteration of this natural environment by a warmer climate, at a time when the world's population is approaching 7 billion, is likely to bring with it a broad array of misery, hardship, and conflict. Precisely how and when these adversities threaten to manifest themselves is discussed later in this chapter. We begin, however, by introducing the organization rightly or wrongly viewed as the gold standard on climate science.

THE INTERGOVERNMENTAL PANEL ON CLIMATE CHANGE

Throughout this book, and particularly in this chapter, there are references to the assessment reports of an entity known as the Intergovernmental Panel on Climate Change (or the IPCC). The IPCC is an organization created under the auspices of the United Nations to advise the world's policymakers on the current state of scientific knowledge on climate change. The IPCC is comprised of approximately 2,500 scientists from over 130 countries. These scientists do not conduct research, but rather are charged with synthesizing the existing peer-reviewed scientific literature and identifying areas of consensus.

The IPCC was established in 1988, and since then it has issued four major assessment reports, in 1990, 1995, 2001, and 2007. The Fifth Assessment Report is scheduled for completion in 2014. The IPCC's three working groups are charged with evaluating what we know about climate change (Working Group I), its likely consequences (Working Group II), and ways to mitigate those potential effects (Working Group III). Each working group selects its own editors, lead authors, and contributing authors. The reports then become standard reference material for government officials, scientists, and the media on climate change issues.*

The IPCC process is far from perfect. First, because of the sheer number of scientists involved and the ponderous nature of their deliberations, it takes several years to complete an assessment. Even before they are printed, the reports are out of date in important respects and are best viewed as snapshots of the state of the science at a particular moment in time. Second, the scientists (and the officials who choose the scientists) are not immune from personal or nationalistic biases. It is probably more than just coincidence, for example, that scientists from Saudi Arabia (with its dependency on fossil fuel production) and China (with its booming coal economy) were among the most uncertain concerning the link between increased global temperatures and human generated emissions.[45] Third, because of the overwhelming consensus among IPCC scientists that climate change presents a clear and present danger, opposing viewpoints can be chilled. In Chapter 6, we consider one example of how the IPCC's herd mentality can stifle opposing viewpoints, the so-called Hockey Stick Graph Controversy.

---

* For example, the U.S. Environmental Protection Agency's findings, in April 2009 that GHGs were "air pollutants" under the Clean Air Act was based heavily upon the IPCC's Fourth Assessment Report.

Despite these flaws, however, IPCC reports can be a very valuable resource. While leaner organizations are capable of producing more contemporary information and cutting-edge theories, the laborious nature of the IPCC offers greater stability and balance. As one IPCC member commented: "The IPCC-like process is the worst way to compile scientific knowledge, except for all the others."*

LETHAL HEAT

The IPCC's Fourth Assessment Report (2007) projected twenty-first-century temperatures by synthesizing the results produced by several leading climate modeling centers from around the world. The IPCC incorporated six different GHG emission scenarios into the models. The models' best estimates under the respective GHG emissions scenarios projected temperature increases ranging from 3.2° to 7.2°F (1.8° to 4°C) over current (1980 to 1999) levels by 2090. The models projected a best-estimate average temperature increase of about 1.8°F (1°C) over 2000 levels by 2045.[†]

Most assessments, including the IPCC reports, conclude that higher GHG levels will also result in greater overall climate variability.[46] The combination of higher average temperatures and increased temperature variability would combine to cause unprecedented heat waves (see Figure 7). The European heat wave of 2003 demonstrates how sustained spikes in temperature can have devastating effects on human populations. As a result of the sweltering heat during August 2003, hospitals were overrun, power went out, crops failed, and tens of thousands died.

---

* John Christie, "No Consensus on IPCC's Level of Ignorance," *BBC News,* November 13, 2007. Christie is viewed in some circles as a contrarian because he has been openly critical of scientists who make catastrophic predictions or steer findings to facilitate a desired political end.

† While the IPCC temperature projections from 2007 remain the benchmark most often cited, more recent reports have been published. For example, in January 2009, an MIT study projected a median estimated temperature rise of 3.1° F (1.9° C) by 2045. *See* A.P. Sokolov, et. al., "Probabilistic Forecast for 21st Century Climate Based on Uncertainties in Emissions (without Policy)" and "Climate Parameters," MIT Joint Program on the Science and Policy of Global Change, January 2009.

There is, however, more to the story than just scorching summers. As scary as "lethal heat" and "extreme spikes in temperature" sound, it is likely that warmer temperatures would result in

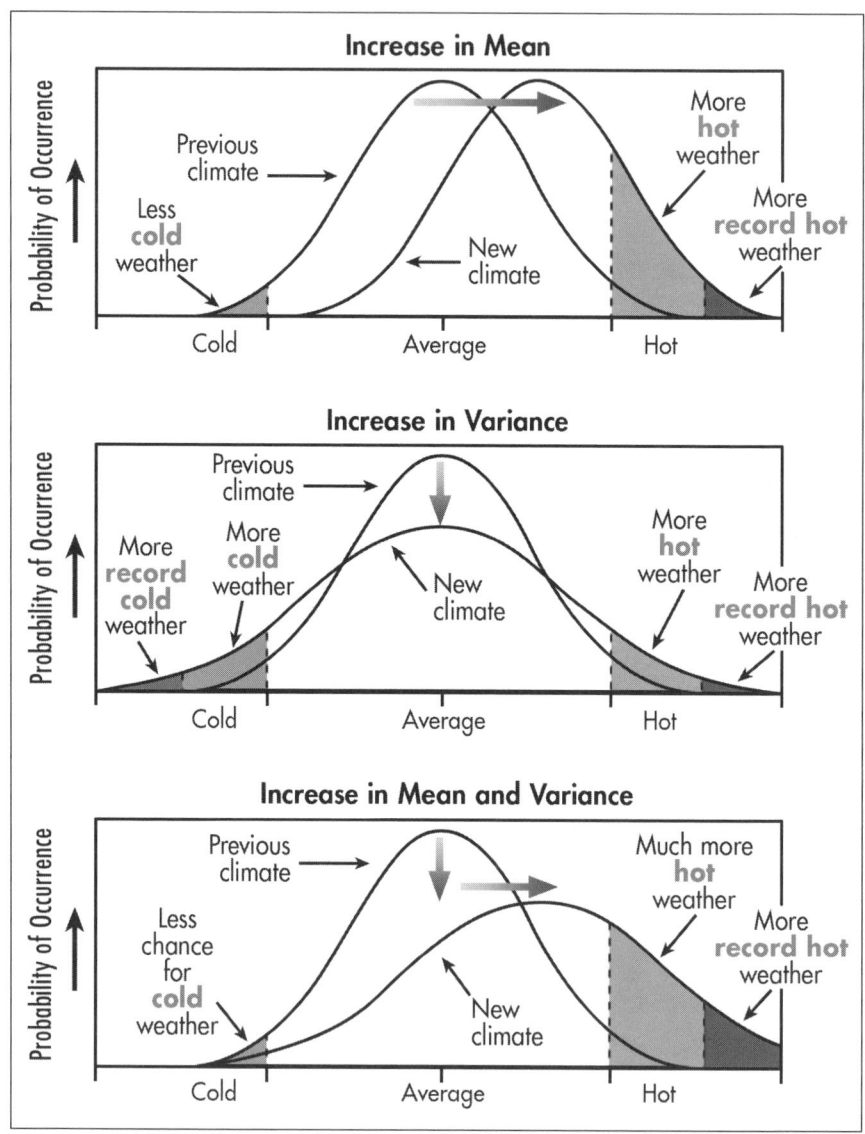

FIGURE 7. Combined effects of increases in temperature mean and variability (*Courtesy of IPCC*)

a net savings of lives.* Cold-related deaths are far more numerous than heat-related fatalities in every area of the world except the tropics.[47] Because more frequent heat waves would also be accompanied by a decrease in bitter cold spells, no analysis of one set of effects is complete without consideration of the other. According to controversial environmental writer Bjorn Lomborg, "By 2050, there will be almost 400,000 more heat-related deaths a year, and almost 1.8 million fewer cold-related deaths. Warmer temperatures will save 1.4 million lives each year."[48]

> **Bjorn Lomborg**, a young Danish academic, is one of the more interesting characters in the climate change debate. As the *Wall Street Journal*'s Keith Johnson wrote: "If the global warming circus has a bad boy it's Bjorn Lomborg." In his book, *Cool It*, Lomborg acknowledges the existence of man-made global warming, but argues that: (1) its impacts have been wildly exaggerated; and (2) proposed cap and trade solutions, while well intended, will not benefit human prosperity. In addition to his writings, Lomborg is known for successfully feuding with Denmark's Committee on Scientific Dishonesty.

Thus, on balance, it is hard to become overly concerned about the direct impact of heat waves upon human health. Instead, as discussed in the sections that follow, it is the potential indirect effects of sustained warmer temperatures upon the world's ecosystems to which the planet's species (including human beings) have adapted, that poses the greatest threat.

RISING SEAS

Over the last 130 years, sea levels have risen about 8 inches (200 mm) as a result of increased temperatures (Figure 8). On average,

---

* This "net savings" refers to lives impacted directly by temperature extremes, and does not take into consideration loss of life caused by migrations, wars, or other potential indirect effects of a changing climate.

oceans rose .062 inches (1.6 mm) per year. While that increase alone is not insignificant, the much greater concern is that the rate of ocean expansion seems to be accelerating.

Worldwide, approximately 150 million people live at an altitude less than three feet above high tide. For island nations sitting on coral atolls just feet above the ocean, such as Tuvalu (in the Pacific Ocean) and Maldives (in the Indian Ocean), rising sea levels present a realistic threat to their very existence. Maldives, which derives significant revenue from tourism, has established a "sovereign wealth fund" to insure against the possibility that its entire nation will have to relocate. Tuvalu, which lacks Maldives's financial resources, has fewer options. As expressed by its prime minister, Saufatu Sopoanga, in an address to the 58th Session of the United Nations General Assembly in New York on September 24, 2003: "We live in constant fear of the adverse impacts of climate change. For a coral atoll nation, sea level rise and more severe weather events loom as a growing threat to our entire population. The threat is real and serious, and is of no difference to a slow and insidious form of terrorism against us."

For island nations, and nations with low-lying population centers, the critical issue is not whether the oceans will rise in the coming decades, but rather how fast they will rise. In the short

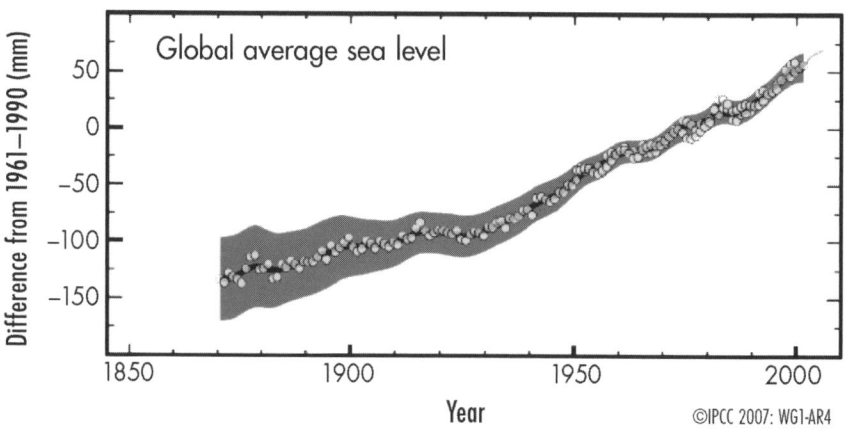

FIGURE 8. Global Average Sea Level *(Courtesy of IPCC)*

term, sea levels can be influenced by a variety of causes, including storm activity, ocean oscillation patterns (such as El Niño), tides, earthquakes, and seasonal variations. In the long term, however, there are just two major influences on the Earth's sea level. First, through a process known as thermal expansion, warmer ocean temperatures expand the volume of the seas. Second, water stored on land, in the form of lakes, rivers, glaciers, and ice sheets,* reduces sea level. Conversely, when water locked away in these forms escapes to the sea, oceans rise. Both thermal expansion and the release of land-locked water are influenced directly by climate change.

The link between rising average world temperatures and thermal expansion of the oceans is direct and well understood. For any hypothetical increase in temperature, we can compute, with reasonable certainty, a corresponding rise in sea level. Thus, once the IPCC agreed upon a range of projected temperature increase for the twenty-first century in its Fourth Assessment Report, it was able to project the thermal expansion that would simultaneously occur during that period.

The relationship between rising temperatures and the loss of land surface ice to the oceans, at least in the short term, is much more complex. Each year, enough snow falls upon global ice sheets to reduce ocean levels by 8 mm. If an equal amount of water returns to the ocean through melting, balance is achieved. When global temperatures rise, there is more melting, but there is also more snowfall. How these countervailing factors interact is the subject of great uncertainty and vigorous debate. In its Fourth Assessment Report, the IPCC projected slight annual decreases in surface ice (i.e., increases in ocean height), but conceded that the response of surface ice to warming is still very uncertain.

For good reason, most discussion of catastrophic coastal flooding centers upon the Earth's two massive ice sheets: Greenland and Antarctica. Melting sea ice does not affect ocean levels because it

---

* "Ice sheets" (larger) and "ice caps" (smaller) are bodies of ice covering a land area. A "glacier" is a large slow moving mass of ice.

displaces its own volume. Melting of all of the world's glaciers and ice caps would result in an ocean rise of less than three feet.[49] The Greenland and Antarctic Ice Sheets, however, hold enough water to raise oceans by 24 feet and 186 feet, respectively. And while there is virtually no evidence that all or most of that ice will melt during this century, there are ample reasons to keep a very close eye on these ice sheets. First, there have been several periods in the last twenty thousand years when melting ice sheets contributed to sea level rise at a rate that was three to ten times faster than the current rate.[50] Second, several new reports suggest that melt-water within the Greenland Ice Sheet may be increasing the flow of land-locked water to the Atlantic.[51]

The IPCC's Fourth Assessment Report projects an average annual rise in sea level—from thermal expansion and loss of land ice combined—of 1.8 to 5.8 mm over the next century. That would mean a total increase of between 180 to 580 mm, or roughly 8 to 23 inches, enough to put many islands underwater and flood low-lying coasts worldwide. Given these vast uncertainties and potentially catastrophic risks, ocean levels and ice sheet volume will be carefully monitored in coming years. If sea level increases begin to trend above the average annual rate of 3.1 mm seen over the last decade (1993 to 2003), it will portend very, very serious consequences.

CHANGING PRECIPITATION PATTERNS: DROUGHT, FLOODS, AND WILDFIRE

Any significant changes to global precipitation patterns would wreak havoc on human civilizations. A delicate equilibrium now exists between complex societies and the water that sustains them. Too much or too little rain over a sustained period of time would upset that balance, leading to different combinations of drought, flood, famine, and wildfire around the world.

There is a strong consensus that warmer atmospheric temperatures will lead to more evaporation of seawater, which, in turn, will result in greater rainfall. A 2007 NASA-supported study concluded that global rain increased at a rate of 1.4% over the two decades between 1987 and 2006.[52] More rainfall, especially in combination with rising ocean levels, increases the potential for floods.

Unfortunately, challenges posed by increases in overall precipitation will be compounded by shifts in existing precipitation patterns. Weather is influenced by air circulation that is caused by disparities in temperatures at the Earth's different latitudes. Because the greenhouse effect adds uneven warmth to the planet (akin to a blanket thicker at some latitudes than at others), an increased greenhouse effect will cause greater temperature disparities—and thus, will redefine the world's atmospheric circulation patterns. In short, global warming will mean more rain and wind in some locations, and less in others.

Although there is substantial agreement that warmer temperatures will cause both drought and flooding, the severity of these anticipated events remains unknown. A 2007 report prepared by the U.S. Government Accountability Office on financial risks posed by climate change concluded that "rising temperatures are expected to increase the frequency and severity of damaging weather-related events, such as flooding or drought, although the timing and magnitude are as yet undetermined."[53] While it is not possible to establish a cause-and-effect relationship between climate change and any particular flood or drought, the devastating "once every one thousand years" Australian drought epitomizes the type of event that the models have predicted.[54]

In addition to altering precipitation cycles, global warming contributes to drought by melting mountain snow earlier in the year. When the thaw occurs gradually, snow protects soil and vegetation for a longer period, and the meltwater provides a steady flow for rivers and streams. When the melt is accelerated, the dry season is extended and forests become tinderboxes. According to a 2006 article published in *Science*,[55] there has been an abrupt increase in the frequency and duration of wildfire activity in the western United States since the mid-1980s. The authors noted that this increase in wildfire activity correlated with a "shift toward unusually warm springs, longer summer dry seasons, drier vegetation (which provoked more and longer burning large wildfires), and longer fire seasons."

EFFECTS ON PLANTS AND ANIMALS

As temperatures increase, animals and plant species will move to cooler latitudes and higher altitudes—unless they cannot. Migration toward cooler locations can be blocked by any number of factors, including natural barriers such as mountains, deserts, and oceans; human alterations to the landscape such as major highways and fencing; or predators. In any case, migration to cooler climates ends at the top of the mountain or the end of the Earth.

The plight of the polar bear is the most highly publicized example of a species that simply has no place left to go. The relationship between melting sea ice and polar bear population was succinctly defined by the U.S. Department of the Interior's Fish and Wildlife Service as follows:

> Although some females will use snow dens on land for birthing cubs, polar bears are almost completely dependent upon Arctic sea-ice for survival. They use sea ice as a platform from which to hunt and feed upon seals, to seek mates and breed, to move to maternity denning areas on land, and to travel long distances. Thus any significant changes in the abundance, distribution, or existence of sea ice would have profound effects at all stages of the animal's life cycle.[56]

As discussed in Chapter 2, the prospect of an ice-free Arctic by the summer of 2020 is no longer a fringe view. According to the Union of Concerned Scientists,* "if there is almost complete loss of summer sea-ice polar bears may not survive as a species."[57]

While much of the media attention has been focused upon the

---

* The Union of Concerned Scientists is comprised of 250,000 scientists and citizens seeking a healthy environment and safer world. Although more moderate and reserved than many environmental groups, aspects of this organization can be fairly described as left leaning by U.S. standards. For example, the UCS supported the Kyoto Protocol at a time when it was overwhelmingly opposed by the U.S. Senate.

endangerment of species that cannot effectively migrate, the more formidable problem is likely to be those plants and animals that *do* migrate. Individual ecosystems consist of a delicate balance of diverse relationships developed over thousands of years. Although constant evolution of ecosystems is a natural phenomenon, there is a growing concern that changes are currently occurring too fast for many species to develop adequate adaptation strategies.

When a plant or animal species migrates, the balance of two ecosystems becomes altered: the one from which a plant or animal departs and the one to which it arrives. Sometimes these comings and goings have little impact, but in other instances, there are major consequences. For example, when warmer average temperatures on the North American continent pushed warblers (small song birds) northward from the United States to Canada, the budworms they had consumed were left unbridled to devour balsam firs.[58] And, warmer temperatures have enabled mountain pine beetles to move north into British Columbia, Canada, where, within the last decade, they have adversely impacted a forest area the size of England.[59]

Over the past century, temperatures have increased 1°F (.6°C), and the impact upon plants and wildlife has been wide-ranging. In February 2009, the Audubon Society issued a report on bird migration changes over the last forty years. According to that report, "Significant northward movement occurred among 58% of the observed species—177 of 305," and "the average distance moved by all studied species—including those that did not reflect the trend—was 35 miles northward."[60] Similarly, the IPCC reported the general trend toward "poleward and upward shifts in ranges in plant and animal species."[61]

If, between 2009 and 2045, temperatures increase an additional 3.1°F (1.9°C) as projected in a 2009 MIT study,[62] plant and animal migrations, and the consequential destabilization of habitats, are likely to get far worse. The changes in precipitation patterns discussed previously will further stress the ecosystems around which humans have built their societies. While rolling the dice does not always result in snake eyes, volatility of this magnitude has the potential to cause untold hardship and ultimately spur international conflict. We consider this latter possibility in the next section.

WHAT IS THE LIKELIHOOD THAT CLIMATE CHANGE WILL LEAD TO MILITARY CONFLICT?

It is almost inevitable that substantial changes to ecosystems and corresponding changes in access to scarce natural resources will spur conflict. Securing access to vital resources is among the oldest and most common causes of human warfare. By way of example, consider why, while already engaged in one expensive conflict, the Japanese would attack Pearl Harbor in December 1941. The answer lies in understanding Japan's dwindling access to critical resources. After the Japanese occupation of Indochina in 1940 and 1941, the United States and other Western nations imposed crippling embargos on iron, oil, and steel. Faced with the prospect of losing 90% of its oil supply, the Japanese did the otherwise unthinkable: It attacked a nation with a Gross National Product six times its own.

The first Gulf War (and, by many assessments, the second) were fought over oil access. Hitler's attack of the Soviet Union in 1941 was motivated by Germany's need for oil from Russian Caucasus and wheat from the Ukraine. Spanish conquistadors fought the Aztecs over gold. And many of the great colonial conflicts, such as the French and Indian War and the Boer War, were fought over timber, fur, gold, and the like.

A particularly troubling, present-day climate-related resource conflict involves the water rights of India and Pakistan. The agriculture of both nations depends on a web of glacier-fed rivers that flow from the Himalayan Mountains. For the most part, these rivers of the Indus System travel through India before reaching Pakistan. As warming temperatures melt the glacier (as is now occurring at a rate of 7% per year), the rivers will swell for a period and then begin to recede. The scarcity of flowing water will ultimately result in a greater proportional use by India, and a life-and-death struggle for many Pakistanis. The governments of these nations, both of which have access to a nuclear arsenal, will face immense pressure to protect access to this dwindling, life-sustaining, water supply.

If the IPCC's most recent projections on climate change are accurate, the world will soon experience increased regional famine, water shortages, and associated mass migration. As one public research institute recently summarized, these changes will "seriously exacerbate already marginal living standards in many Asian, African, and Middle Eastern nations, causing widespread political instability and the likelihood of failed states. . . . The chaos that results can be an incubator of civil strife, genocide, and the growth of terrorism."[63]

The conflict in Darfur is often cited as an example of climate change–induced warfare. While it is overly simplistic to neatly package genocide as yet another symptom of global warming, a strong case of correlation can be made. As a result of the Indian Ocean's increased average temperatures, rainfall has declined 40% over the last two decades in parts of sub-Saharan Africa. Africa's largest nation, the Sudan, and in particular its Darfur region in the west, has experienced extreme drought. When water was more plentiful, Darfur's farmers shared it with nomadic herders of that region. By 2003, however, violence erupted when the African farmers sought to protect their land from the Arab pastoralists. That conflict, compounded by governmental mismanagement, cascaded into the death, rape, carnage, and displacement of entire farming communities. In a June 2007 *Washington Post* editorial, U.N. Secretary General Ban Ki Moon wrote, "Amid the diverse social and political causes, the Darfur conflict began as an ecological crisis, arising at least in part from climate change."[64]

Finally, it is not just suffering and shortages that could lead to armed conflict, but also competition over new resource opportunities. With the melting of polar ice caps, and the opening of the Northwest Passage in 2007, for the first time in recorded human history it has become possible to circumnavigate the North Pole. In what some environmentalists view as a bitter irony, this GHG-induced Arctic melting has created access to an abundance of previously inaccessible fossil fuels. According to the U.S. Geological Survey, as much as one quarter of the world's undiscovered oil and gas reserves are in the Arctic,[65] under areas that once held year-round ice. The race to claim this great energy prize has become highly contentious.

In August 2007, two Russian mini-submarines planted their nation's flag on the seabed 14,000 feet below the North Pole. Canada's minister of foreign affairs, Peter MacKay, offered the following chilly response: "This isn't the fifteenth century . . . you can't go around the world and just plant flags and say 'We're claiming this territory.'" Canadian prime minister Stephen Harper announced that ships entering the Northwest Passage should first report to the Canadian government. A 1982 U.N. agreement known as the Convention on the Law of the Sea will ultimately play a significant role in this boundary dispute between the Arctic Nations (Canada, Russia, Denmark/Greenland, Norway, and the United States) but it is important to note that in addition to diplomacy and sharp debate over international law, the Arctic region has seen a marked increase in the scope and frequency of military exercises.[66]

CLIMATE CHANGE AND DISEASE

Most lists projecting the dangers of global warming include increased exposure to vectorborne disease. A vectorborne disease is one in which a virus or other infectious agent is transmitted from individual to individual, or from animal to individual, by mosquitoes, flies, ticks, or some other vector. Vectorborne diseases, such as malaria, dengue fever, and yellow fever, are much more prevalent in tropical and subtropical locations because vector insects thrive in those climates. Many scientists are concerned that, as surface conditions become warmer and more humid, the range of mosquitoes and other vectors will expand—and that these diseases will follow them, occurring with greater frequency, further from the equator, and at higher altitudes.

Over the last decade, there have been persistent reports of vectorborne diseases spreading to new regions as a result of climate change. For example, in May 2006, the *Washington Post* reported: "Global warming—with an accompanying rise in floods and droughts—is fueling the spread of epidemics in areas unprepared for the diseases, say many health experts worldwide. Mosquitoes, ticks, mice and other carriers are surviving warmer winters and expanding their range, bringing health threats with them."[67]

Although there is an abundance of scientific opinion linking warmer weather to infectious disease, many of the major scientific bodies have remained noncommittal on this issue. A 2001 study by the U.S. National Academy of Sciences concluded that, while there is an "obvious and long appreciated" relationship between climate and human health, there is "little solid scientific evidence" to link recent changes in infectious disease patterns with climate change.[68] The IPCC (Fourth Assessment) in 2007 was also guarded in its conclusions on vectorborne disease, noting only that "the spatial distribution of some infectious disease" may have been "altered" by global warming.[69]

The real risk of vectorborne disease, however, is not what scientists can now prove will happen, but rather what they believe may happen. Environmental changes are often accompanied by the appearance of new diseases, and warmer temperatures drive such environmental change. As one commentator noted, "Climate change will shuffle the deck of plants, animals, and ecosystems in ways we've only begun to imagine."[70] The risk that those changes will lead to a surge of new exotic diseases, a resurgence of known diseases, or both, is one of many that must be weighed against the costs of reducing GHG emissions.

MORE INTENSE HURRICANES AND STORMS

Hurricanes are among the most powerful and destructive forces on Earth. They can extend 600 miles across, reach speeds of 200 mph, and devastate entire coastal communities. Fortunately, the convergence of conditions necessary to form hurricanes is relatively rare, and exists only in limited geographic locations. When ocean temperatures reach 80°F (26.5°C), powerful storms circulating in the atmosphere can create a tropical depression. When prevailing wind conditions cause the storms to rotate, the rotation creates a vortex that pulls warm, wet ocean surface air upward, where it releases its heat energy and generates high winds. As the air beneath the vortex rises, a low-pressure area is created near the surface, causing additional humid air to rush in. In effect, the humid air fuels the vortex, enabling the vortex to pull in more humid air. At 39 mph (34 kts, 17.5 m/s), these circulating winds constitute a tropical storm, and when they reach 74 mph (64 kts, 33m/s), a hurricane has formed.[*]

Climate change impacts hurricane activity[†] in at least three ways. First, higher ocean temperatures provide a more abundant source of hurricane "fuel." As hurricanes expand, cooler water is welled up from farther beneath the surface into the vortex, which has a literal chilling effect on the storm's intensity. But because global warming impacts water at depths down to 1,500 feet, this natural safety valve may be waning.[71] Second, climate change has increased the amount of water vapor along ocean surfaces. Like higher ocean temperatures, more ambient water vapor will theoretically fuel more frequent and more intense hurricane activity. And third, some climate models suggest that increased wind shear (i.e., changes in wind direction or speed at different levels of the atmosphere) caused by a warmer climate will have the opposite effect, by inhibiting the development or intensification of hurricanes."[72]

While there is general agreement that warmer waters and more humid air have the capacity to fuel hurricane activity, there is also a consensus that more wind shear will suppress hurricane activity. There is great debate, however, as to which of these countervailing forces will ultimately dominate the other. The IPCC (Fourth Assessment) concluded that: (1) it was "more likely than not" (more than a 50% probability) that warming had increased the frequency of North Atlantic hurricanes; and (2) it was likely (more than a 66%

---

[*] Generally, a weather event found in the North Atlantic and Eastern Pacific regions, similar occurrences are referred to as "typhoons," "severe tropical cyclones," and "severe cyclonic" storms in other parts of the world.

[†] "Hurricane activity" is measured in terms of hurricane frequency and hurricane intensity.

probability) that warming would result in more intense hurricanes, typhoons, and tropical cyclones in the future.[73]

Shortly after the 2007 IPCC report's release, several other reports were unveiled casting further doubt on the relationship between climate change and hurricane activity. Two separate evaluations by NOAA-affiliated scientists, each using new modeling techniques to assess hurricane formation, concluded that a robust increase in wind shear created by higher temperatures could counteract the effects of warmer seas.[74] However, the biggest blow to the theory that warmer waters will result in greater hurricane activity was delivered from an unlikely source. In March 2008, one of the titans of hurricane research, Kerry Emanuel of MIT, concluded that (according to a new modeling technique) global warming should reduce the frequency of worldwide typhoons and hurricanes.[75] This revelation is particularly remarkable because Emanuel had theretofore been a highly visible proponent of the view that warmer oceans could fuel an explosion of powerful storms. Indeed, Emanuel's influence was such that *Time* had previously designated him as one of its *100 Most Influential Persons of 2006*, because his work represented an important "trigger point" in convincing Americans that the Earth is growing warmer.[76]

As with many of the potential manifestations of climate change, our understanding of the relationship between warmer global temperatures and hurricanes remains unsophisticated. Records of cyclonic activity before 1970 are not completely reliable, and models yield inconsistent results. In terms of consensus, therefore, the most that can be said is that: (1) we have little idea as to how global warming will affect the frequency of hurricanes and (2) there is good reason to believe that warmer oceans (and increased surface humidity) have the capacity to intensify hurricanes.

In these first five chapters, we have examined the theory, evidence, and dangers of global warming. In an effort to provide a balanced discussion, these pages have been sprinkled with nonconsensus views. Next, however, we venture directly into the strongholds of skepticism to see what might be learned from those who offer dissenting opinions.

6

Climate Change Skeptics
Scientific Champions or Modern-Day Charlatans?

"Fifteen percent of the population believe the moon landing was actually staged in a movie lot in Arizona, and somewhat fewer still believe the Earth is flat. I think they all get together with the global warming deniers on a Saturday night and party."
—AL GORE (SEPTEMBER 2006)

"It is difficult to get a man to understand something when his salary depends upon his not understanding it."
—UPTON SINCLAIR (1935)

Too often, those who offer an unconventional view on climate change are reflexively marginalized as part of a fringe element or described as financially (or ideologically) motivated. While these characterizations certainly befit some skeptics, there are many distinguished scientists who take issue with one or more aspect of the consensus view on climate change. This chapter examines several of the views outside of the mainstream. Rather than merely seek to set these arguments up and knock them down like straw men, these alternative viewpoints are given fair consideration. We begin by acknowledging the important historical role of skepticism in science.

THE HONORABLE TRADITION OF SKEPTICISM

Skepticism is the lifeblood of science, critical to any meaningful scientific advancement. It is the intense questioning of conventional wisdom that opens the door for new theories and advancement. As American sociologist Robert Merton put it: "Most institutions demand unqualified faith, but the institution of science makes skepticism a virtue."[77] Healthy skepticism and the courage of independent conviction are the last lines of defense against the powerful and dangerous force of "groupthink."*

There is a fine line, however, between skepticism and cynicism. While a skeptic demands reason or evidence, a cynic refuses to believe, regardless of reason or evidence. As one author put it: "Cynicism and gullibility together produce a penchant for magical thinking and the suspension of logic. Cynicism produces disdain for the traditional methods and sources of information; gullibility leads us to embrace idiosyncratic ones instead. Charlatans and opportunists have been quick to take advantage of these traits."[78] For example, many of the scientists and policymakers refusing to acknowledge the dangers of asbestos and tobacco—in the face of overwhelming evidence—were cynics posing as skeptics. The trick then, is to distinguish honest skepticism from cynicism and propaganda. However, that is easier said than done. Is a scientist funded by a conservative think tank or the fossil fuel industry necessarily a shill or a stooge? If so, what does that say about "mainstream" scientists who only have a job because a consensus exists that anthropogenic global warming is occurring? The best analytical approach would seem to be two-fold: (1) take a moment each time we read an article, glimpse a billboard, or watch a television report, and question the source; and (2) recognize that climate science is still in its formative stages and keep an open mind to new ideas, unexpected results, and novel theories. The section that follows provides an example of what can happen when non-consensus views are dismissed out of hand.

---

* The phenomenon of groupthink was described by psychologist Irving Janis, who researched it extensively as: "A mode of thinking that people engage in when they are deeply involved in a cohesive in-group, when the members' strivings for unanimity override their motivation to realistically appraise alternative courses of action."

Cautionary Tale Number One:
The Great Hockey Stick Graph Controversy

The greatest icon of the climate change debate is the aptly named Hockey Stick Graph (Figure 9). This simple illustration depicts relatively flat climactic conditions for a millennium (the handle shaft), followed by a dramatically sharp rise in temperatures over the last century (the blade). If a picture is worth a thousand words, this graph is worth a million. At a glance, it seems to demonstrate that the combustion of fossil fuels is drastically altering the climate. According to its critics, however, it is a hoax—a product of bad science, a result of herd mentality, and evidence of IPCC bias.

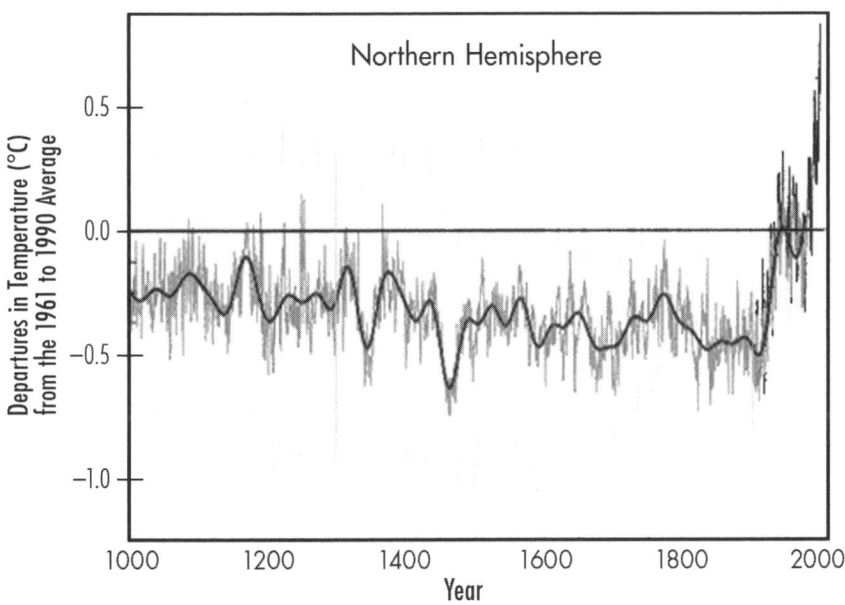

FIGURE 9. The Hockey Stick Graph *(Courtesy of the IPCC)*

The story of the Hockey Stick Graph begins in the late 1990s. By then, the Kyoto Protocol had been negotiated and climate change had become a mainstream issue. Information on historical temperatures, however, remained sketchy, and direct temperature records were only available from around 1860. Beyond that, there was an assortment of proxy measurements taken from tree rings, as well as from ice core and coral isotopes.* A team led by American climatologist Michael Mann constructed a statistical model designed to digest the available grab bag of proxy measurements, and to yield a comprehensive reconstructed 1,000-year temperature record. The Hockey Stick Graph is a visual representation of the Mann team's conclusions (although Mann's gray shaded margin of error is sometimes removed by others for effect).

Mann's results were eagerly accepted by most of the scientific community as further proof that anthropogenic global warming was well underway. Indeed, in its 2001 Summary for Policymakers, the IPCC prominently featured the Hockey Stick Graph as evidence that "[i]t is also likely that, in the Northern Hemisphere, the 1990s was the warmest decade and 1998 the warmest year [over the last millennium]."[79] Prior to Mann's work, it had been largely accepted that the Medieval Warm Period (A.D. 800–1300) and the Little Ice Age (A.D. 1650–1850) had represented significant global climatic events.[80] Many scientists who did not share the consensus view on global warming were shocked that, despite the Mann team's refusal to disclose its statistical model, its conclusions had been presented to the world's most influential policymakers (i.e., the IPCC) as accepted science. As summed up by one of those scientists, the late John Daly:

> In every other science when such a drastic revision of previously accepted knowledge is promulgated, there is considerable debate and initial skepticism, the new theory facing a gauntlet of criticism and intense review. Only if a new idea survives that process does it become broadly accepted by the scientific peer group and the public at large. This never happened with Mann's "Hockey Stick." The coup was total, bloodless, and swift as Mann's paper was greeted with a cho-

---

* Besides being the name of minor league baseball teams in both Albuquerque and Springfield (the one on the *Simpsons*), isotopes are atoms with extra, or missing, neutrons. Isotope relationships provide insights into historical temperatures.

rus of uncritical approval from the greenhouse industry. Within the space of only twelve months, the theory had become entrenched as a new orthodoxy.[81]

What followed next was the scientific version of the Hatfields and the McCoys. Beginning in 2003, two Canadians, statistical mathematician Stephen McIntyre and economist Ross McKitrick, published articles that were highly critical of Mann's techniques and input data. Others joined the criticism, asserting either that Mann's proxy data was flawed or that it had been statistically manipulated. The debate became heated, but because it concerned rather obscure statistical arguments, it was largely confined to the scientific community. That all changed, however, in 2005, when Congress decided to enter the fray.

Texas Congressman Joe Barton, the Republican Chair of the House Energy Committee, requested that a team of nationally renowned statisticians, led by Edward Wegman, review Mann's analysis (the Wegman Committee). Shortly thereafter, Congressman Sherwood Boehlert, the Republican Chairman of the House Energy and Commerce Committee, asked the National Academy of Sciences to evaluate and explain the available scientific evidence on temperatures over the last two thousand years (the NAS Committee).

While the NAS panel members concluded that there were some statistical shortcomings in Mann's analysis and that some of his conclusions could not be justified, they also stated that "the basic conclusion of Mann . . . has subsequently been supported by an array of evidence."[82] While the NAS Report has been characterized as both "an exoneration" of Mann's team's work on the one hand, and a complete invalidation of its conclusions on the other, in truth it was neither.

The Wegman Report, which focused primarily on statistical analysis, was much more unkind to Mann and his colleagues. It concluded that Mann's findings were "somewhat obscure and incomplete," and that McIntyre and McKitrick's criticisms were "valid and compelling." The report then went further, suggesting that the existence of "a social network of the 75 most frequently published authors in the area of climate reconstruction" may have compromised their independence. Shockingly, Wegman's team had called out leading paleoclimatologists, suggesting that they had unwittingly engaged in the great scientific taboo of groupthink. Some of this seemed particularly unfair to Mann, since his original 1998 article had stressed "uncertainty and limitations" in the proxy data.

So is the Hockey Stick saga the commendable story of scientific pioneers making important, if somewhat imperfect, progress? Or, is it a cautionary tale of a mainstream international science community that, in its haste to support a "preferred" conclusion, was willing to overlook established protocol and methodology? Possibly, it is both.

Cautionary Tale Number Two: Global Cooling

On April 28, 1975, *Newsweek* published a story on climate change that was, in many striking respects, similar to contemporary articles on global warming.[83] The story stated:

> There are ominous signs that the Earth's weather patterns have begun to change dramatically and that these changes may portend a drastic decline in food production—with serious political implications for just about every nation on Earth. . . . The evidence in support of these predictions has now begun to accumulate so massively that meteorologists are hard-pressed to keep up with it. . . . To scientists these seemingly disparate incidents represent the advance signs of fundamental changes in the world's weather. . . . The longer planners delay, the more difficult will they find it to cope with the climatic change once the results become grim reality.

The *Newsweek* article differed from today's media stories in one critical respect—it asserted that the anticipated drought, famine, tornadoes, temperature extremes, altered precipitation patterns, and human migration would be the result of global *cooling*. And *Newsweek* was not alone. In the mid-1970s, little more than a decade before James Hansen came before Congress to sound the alarm on global warming, respected publications like the *New York Times*, the *Christian Science Monitor, Time,* and *National Geographic* all ran stories regarding the dangers of cooling temperatures.[84]

Today, several skeptical commentators use the global cooling scare of the 1970s to suggest that contemporary concerns about global warming may be overblown.[85] This isn't an argument directed at the science, but rather one that seeks to undermine the credibility of the media and scientists who communicate the science. The skeptics ask rhetorically: "If the media and the scientific community were so spectacularly wrong about global cooling, why should we put our trust in these same institutions when it comes to global warming?"

To evaluate this line of thinking, we must look beyond the headlines and consider the state of affairs when these articles were published. In so doing, we learn that, after a period of moderate cooling from the 1940s to the 1970s, the popular press was engaged in some pretty wild speculation. However, it also turns out that much of the mainstream scientific community at the time actively distanced itself from this hype. In a 1975 report, the National Academy of Sciences (the closest thing to a collective scientific voice in the United States) expressly repudiated any alleged consensus on global cooling. According to the NAS, "Not only are the basic scientific questions [relating to climate change] largely unanswered; but in many cases we do not yet know enough to pose the key questions."[86] In February 2008, a National Climatic Data Center scientist published a survey of peer-reviewed scientific articles published in the late 1960s and 1970s on climate related issues.[87] The survey showed that, among those articles anticipating climate change, global warming predictions exceeded global cooling predictions by more than six to one.

If there is a lesson to be learned from the global cooling scare, it is that some members of the media may be predisposed to engaging in speculative sensationalism. Not much of a news flash there, but certainly a legitimate point to consider. Those skeptics who seek to use this episode to discredit our scientific bodies, however, do little to advance the debate. Rather, the argument—that because a small number of scientists with access to sparse data got it wrong in the 1970s we should discount today's far more sophisticated analyses by the IPCC, NAS, NOAA, and NASA—simply does not hold water.

SOLAR VARIATION

As discussed in Chapter 2, most scientists now believe that increased atmospheric GHGs have been the primary cause of global warming over the last one hundred years. There is, however, growing support for the view that changes in the amount of sunlight reaching the Earth may have also contributed to twentieth-century temperature increases. Those scientists who suggest that solar variation has caused (or contributed to) the current warming trend generally base that assertion on some combination of three related theories: Milankovitch cycles, sunspot activity, and cosmic rays.

Milankovitch cycles explain differences in the levels of solar energy reaching the Earth over large periods of geological time. In 1941, a Serbian engineer named Milutin Milankovic published theories that cyclical variations involving both: (1) how much the Earth tilts on its axis and (2) the Earth's "not quite circular" rotation around the sun, would affect how and where the Earth received sunlight. It is now widely theorized that these cycles (41,000 years for the Earth's axis "wobble" and 21,000 years for its orbital changes) can combine to affect long-term climate and drive ice ages.* While Milankovitch cycles provide fascinating insights into our extended climate, they do little to explain the recent

---

* The 21,000-year orbital cycles are defined by the date that the Earth is closest to the sun (the perihelion). During that cycle, the perihelion rotates around the calendar (it now occurs in early January). Depending upon when the perihelion occurs, different ratios of land and water receive different amounts of direct sunlight. In turn, this 21,000-year-cycle is affected by a 100,000-year cycle in which the "extent of the eccentricity of the earth's orbit" (i.e., how elliptical or "egg-shaped" the orbit is) changes. The more elliptical the rotation becomes, the more the 21,000-year-cycle impacts climate.

increase in average global temperatures. According to NOAA, these astronomical variations "occur over thousands of years, and the climate system may also take thousands of years to respond to an orbital forcing."[88]

In addition to the epochal Milankovitch cycles, solar radiance is also driven by more contemporary eleven-year sunspot cycles (see Figure 10). Sunspots are simply dark patches that are visible on the surface of the sun. While much about sunspots remains unknown, it has been observed that increases in sunspot activity coincide with periods when the sun radiates more heat than usual. That is, the more sunspots, the more active the sun. Sunspot activity generally peaks during the fifth or sixth year of the sunspot cycle.

Climate fluctuations during the eleven-year sunspot cycle appear minor, with temperatures only slightly higher during the solar maximum (when sunspot activity is at its peak) than during the solar minimum. The more compelling significance of sunspot activity appears to be the relationship between temperatures and the overall size of the cycles. In other words, the net solar activity during an eleven-year cycle (or series of cycles) appears to be an important climate-forcing mechanism.

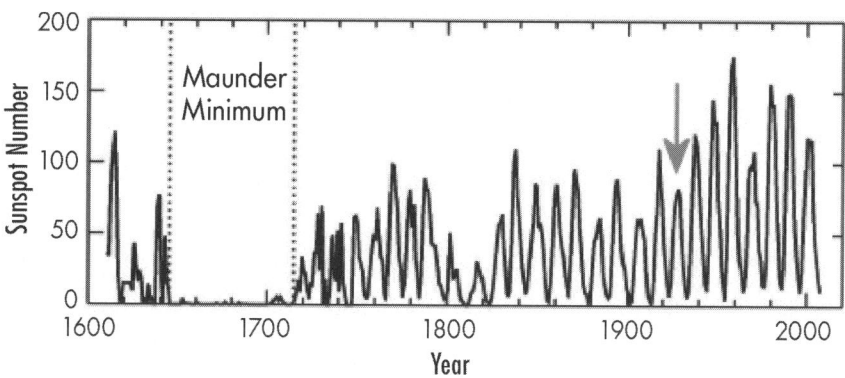

FIGURE 10. Sunspot Activity Over the Last Four Centuries (Courtesy of NASA/MSFC)

During the Maunder (1645–1715) and Dalton (1790–1820) Minimums, when sunspot activity was exceptionally low, the Earth experienced prolonged periods of cooling. Conversely, since the 1950s, a little more than a decade before global temperatures began a steady and steep upward trend, solar cycles have been very intense (See Figure 10).[89] According to a commonly cited 2005 paper cowritten by Russian solar physicist Ilya Usoskin there has been a strong correlation between sunspot activity and global temperatures over the last 1,150 years—with temperatures lagging solar activity by about ten years.[90]

Despite the undeniably intriguing historical correlations between sunspot activity and temperature, the theory that solar variation has driven climate change over the last century has a significant shortcoming. The correlated trend appears to have come to a halt around 1980, when temperatures continued to rise despite a leveling-off of sunspot activity. As stated in the Usoskin article: "During the last 30 years, the solar total irradiance . . . has not shown any significant secular trend, so that at least this most recent warming must have another source."[91]

In addition to gauging the effects of Milankovitch cycles and sunspot activity, climatologists also have to contend with a third solar variant known as cosmic rays. Despite their grandiose name, cosmic rays are simply energy particles that reach the Earth's atmosphere from outer space. Although these particles often originate on the sun, more powerful cosmic rays come from outside of our solar system. An explanation of the theoretical relationship between solar variation, cosmic rays, and temperatures goes something like this: (1) increased solar activity causes increased solar wind; (2) stronger solar wind allows fewer cosmic rays to enter the Earth's atmosphere; (3) a decrease in atmospheric cosmic rays reduces the number of certain particulates that form clouds; and (4) fewer clouds reduce the Earth's albedo (i.e., less solar energy is reflected), causing higher temperatures. The condensed version of this cosmic ray theory is that greater sun activity results in fewer cosmic rays, which drives higher temperatures. Although the mainstream scientific community remains skeptical of the notion that cosmic rays play a major climactic role, there are many scientists counted among the consensus who believe the effect of cosmic rays on clouds is a phenomenon worth exploring.[92]

The world may not have to wait too long for definitive guidance on the issue of whether GHGs, solar variation, or both, drive climate. In 2008 and 2009, sunspot activity was exceptionally low, and NOAA scientists predict that the next eleven-year sunspot cycle (which is expected to peak in 2013) will be mild. In May 2009, the chairman of a NOAA panel on sunspot activity stated: "If our prediction is correct, [the next solar cycle] will have a peak sunspot number of 90, the lowest of any cycle since 1928 when Solar Cycle 16 peaked at 78" (see arrow on Figure 10). With GHG levels increasing rapidly, and sunspot activity on a sharp decline, the proof may soon be in the "climate pudding"—with the trend in temperature revealing which mechanism is the primary climate-forcing agent.

THE CLIMATE IS ALWAYS CHANGING

Long before man exhaled a single breath of carbon dioxide, the world regularly experienced sweeping temperature change. As accurately portrayed in the skeptical *New York Times* best-seller, *Unstoppable Global Warming (Every 1500 Years)*: "The Earth continually warms and cools. The cycle is undeniable, ancient, often abrupt, and global. It is also unstoppable."[93] Just fourteen thousand years ago, the massive Wisconsin Ice Sheet covered a large portion of the northern United States.[94] More recently, the Medieval Warm Period (A.D. 800–1300) and the Little Ice Age (A.D. 1650–1850) presented contrasting climate phases—although substantial dispute exists over whether these were global or regional events.[95] The precise reasons for these non-anthropogenic global temperature changes remain unknown. The most likely causes include orbital variations and solar activity (see previous section), continental drift, volcanic activity (Chapter 4), and natural increases or de- creases in greenhouse gas concentrations.

The "climate is always changing" argument comes in two basic forms. The logic of the first version is unassailable: When evaluating the risks of anthropogenic global warming, it is wise to keep in mind that natural changes occur on annual, decadal, and epic time scales, and that to some degree, the causes of these natural temperature trends remain a mystery. A second, more far-reaching, version of the argument implies that because natural climactic variations have historically occurred, it is safe to assume that the current warming is not man-made. This argument is not logically sound. As summed up by the IPCC in 2007 (Fourth Assessment, Working Group I): "The fact that natural factors caused climate changes in the past does not mean that the current climate change is natural. By analogy, the fact that forest fires have long been caused naturally by lightning strikes does not mean that fires cannot also be caused by a careless camper."[96]

The bottom line is that just because warming and cooling trends can be natural, does not mean that they cannot also be anthropogenic. The key is determining, from limited available data, the extent of the human fingerprint on the warming over the last century. As discussed in Chapter 2, the consensus view is that increased atmospheric GHGs is the best, current explanation of this trend. As also discussed in Chapters 3 and 4, however, many pieces of the Earth's complex climate puzzle remain unrevealed—and it is certainly possible that some not-yet-understood natural phenomenon is also forcing global average temperatures upward.

THE BENEFITS OF WARMING

Temperature increases and precipitation changes of the sort predicted by the IPCC would usher in the "parade of horrible" discussed in Chapters 2, 3, and 5. Yet, could such climactic changes also bring about significant societal benefits? The answer, according to both sides of the debate, is "yes."

The advantages of a warmer, wetter world would be wide-ranging. As a result of increased arable land, more CO_2, and longer growing seasons, there would likely be an increase in global agricultural yields. Melting ice would open up the historically treacherous Northwest Passage sea route that runs through the Canadian archipelago connecting the Atlantic and Pacific Oceans (See Figure 11). On balance, warmer weather would reduce heating costs more than it would increase air conditioning costs. And, as discussed in Chapter 5, moderate global warming would likely result in a net savings of lives.

Those who assert that these many benefits of global warming should be acknowledged before embarking upon a costly mitigation strategy, make an undeniably valid point. Any discussion of lost lives, famine, and higher energy costs associated with climate change, without mention of the health, agricultural, and energy benefits of a moderately warmer world, is incomplete.

Some skeptics push this argument further, however, suggesting that a warmer world will certainly be a better world. They assert, for example, that "archaeological evidence shows that people lived longer, enjoyed better nutrition, and multiplied more rapidly [during warm periods] than during epochs of cold."[97] Conspicuously absent from this analysis is any consideration of how climate instability and unpredictability might impact humanity.

If mankind were colonizing an unpopulated Earth, it is possible that an average global temperature of 60°F (16°C) would be prefer-

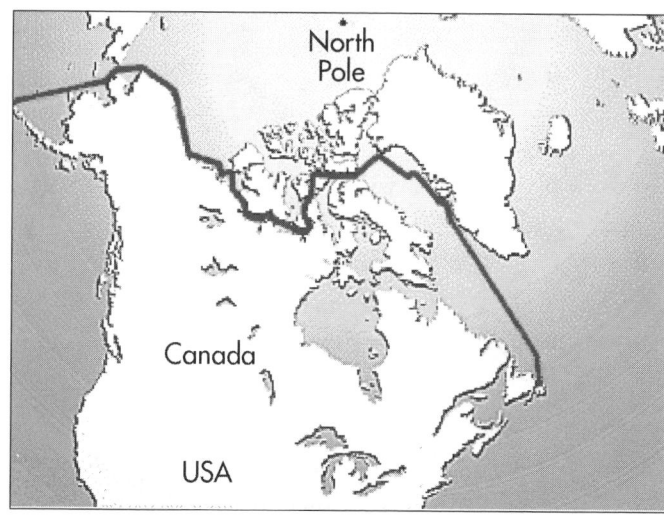

FIGURE 11.
The Northwest
Passage (2009)

able to 58°F (14°C). Similarly, if humans were still hunting and gathering, or if much of the Earth remained unsettled, "warmer is better" might be a concept worth debating. Today's world, however, is one in which 6.7 billion people live in a delicate balance with existing ecosystems. Overflowing population centers depend upon existing river flow, coastlines, and crop patterns to sustain life and to maintain stable international relationships and borders. The commonly expressed argument that a warmer climate assisted the Vikings in colonizing Greenland,[98] and therefore should be good for us, fails to take into account the potential for massive societal destabilization that would almost certainly accompany a reshuffling of the climate deck.

URBAN HEAT ISLANDS

An urban heat island is a highly populated area that is warmer than the surrounding countryside. The U.S. Environmental Protection Agency explains the phenomenon of urban heat island effect (UHIE) as follows: "Heat islands form as vegetation is replaced by asphalt and concrete for roads, buildings, and other structures necessary to accommodate growing populations. These surfaces absorb—rather than reflect—the sun's heat, causing surface temperatures and overall ambient temperatures to rise."[99]

The UHIE can be very substantial. Many cities, and increasingly, suburbs, have temperatures up to 10°F (5.6°C) warmer than the surrounding natural land cover.[100]

The existence of a robust UHIE affect is well-documented and universally accepted. There is vigorous debate, however, concerning whether this effect has influenced global average temperature estimates in such a way as to exaggerate the current warming trend. The skeptical view, as succinctly expressed by Douglas Hoyt, a solar physicist and climatologist, is that "UHI have significant impacts of measured trends . . . about 45% of the observed warming is actually spurious urban warming."[101]

Most climate scientists, as well as the IPCC, take the view that increases in average global temperature are "very unlikely to be influenced significantly by increased urbanization."[102] Several of the bases for the consensus view that the UHIE has not upwardly skewed average global temperatures, as well as the skeptical response to those arguments, follow:

- The temperature statistics used by climate models either exclude or adjust data from anomalous population centers that appear to be influenced by the UHIE—thereby negating any effect. *Skeptics' response: Because more and more locations are experiencing urban and suburban sprawl, it is increasingly difficult to define which locations are indeed "anomalous."*

- Satellite and oceanic readings, unaffected by the UHIE, also demonstrate a strong warming trend. *Skeptics' response: Some studies indicate that temperatures measured by satellites and weather balloons demonstrate a much more modest warming than land surface measurements.[103]*

- Most of the measured warming has occurred in rural areas, with the Arctic being particularly impacted. *Skeptics' response: Focusing on particular rural anomalies, and ignoring others (like the Antarctic), serves little purpose. Since we know substantial UHIE is occurring, why would we assume that it is having no influence on average global temperatures?*

- Since urban temperature trends are not materially different from rural trends, the UHIE hypothesis fails. *Skeptics' response: Development is not unique to urban areas, but rather occurs in suburban and rural areas as well, so it is not surprising that trends are similar in urban and rural areas.*

The bottom line is that there are many compelling reasons to doubt that urban heat islands are responsible for much of the measured warming over the last century. Because the UHIE does have some impact on land surface temperature readings, however unquantifiable, and because land surface records are relied upon by climatologists, this issue is likely to remain a subject of further study and intense debate for many years.

CARBON DIOXIDE IS A "MINOR" GREENHOUSE GAS

In support of the assertion that "carbon dioxide is actually a puny player in the greenhouse game,"[104] skeptics generally point to three facts. First, carbon dioxide is a relatively rare gas in the atmosphere. Even with a 40% increase since the industrial revolution, CO_2 concentration in the ambient air is only 390 parts *per million*. Second, natural processes emit thirty times more CO_2 than do human beings. And third, water vapor—not CO_2—is the most abundant and dominant GHG. While these irrefutable facts seem compelling at face value, they represent only part of the story.

It is true that CO_2 makes up only a small portion of the atmosphere, but it is well-established that, even in limited amounts, this powerful gas has the ability to trap vast amounts of heat. CO_2 is generally estimated to account for between 9 and 26% of the greenhouse gas effect.[105] Thus, despite its low concentration levels, CO_2 plays a major role in the greenhouse process that warms temperatures by 60°F (30°C), and makes the Earth inhabitable (see Chapter 2).

It is also true that anthropogenic CO_2 emissions are just a small percentage (3%) of natural CO_2 emissions. Taken out of context, however, that can be a misleading statistic. Before human beings emitted that extra 3% of CO_2, the Earth was in a general state of carbon equilibrium. For every gigaton of CO_2 that was going into the atmosphere, another gigaton was being cycled out. It has been a different story, however, since humans have altered that equilibrium. Because the Earth's carbon sinks are only able to absorb about half of the additional anthropogenic CO_2, the atmospheric carbon imbalance continues to grow. Over the last two centuries, ambient carbon dioxide has increased from 280 ppm to 390 ppm. There is little doubt in the scientific community, even among skeptics, that it is the man-made emissions that are tipping the CO_2 balance. Thus, the oft-repeated statement—that "man's carbon emissions are only a small percentage of the Earth's natural emissions"—is both technically accurate and largely irrelevant in considering the issue of climate change.

Finally, it is true that water vapor—not carbon dioxide—is the world's most abundant GHG. Water vapor accounts for somewhere between 66 and 85% of the GHG effect,[106] which is three to nine times more than that of CO_2. There are, however, two important considerations that skeptics often leave unmentioned. First, the fact that water vapor traps more heat than CO_2 does not mean that CO_2 is not an important GHG. To the contrary, CO_2's contribution to the greenhouse effect (9 to 26%) is substantial. Second, CO_2 is much more likely to serve as a catalyst for climate change than water vapor. Water vapor has an atmospheric lifespan of about one week. At any given time, the amount of water vapor in the air is simply a function of the temperature, with warmer weather increasing the levels of moisture in the air. Conversely, CO_2 has much more staying power; it can remain in the atmosphere for over a century. Thus, CO_2 levels can be significantly increased by man over time. Increased CO_2 captures more heat, which in turn produces more water vapor. Through this feedback mechanism, the warming effect of CO_2 is further amplified by increased water vapor.

THE LIMITATIONS OF CLIMATE MODELS

Climate models simulate how key climatic factors (like oceans, land, atmosphere, ice, solar energy, and living things) interact under various prescribed conditions. These ever-increasingly complex models provide some of the most compelling evidence for the case that immediate, decisive mitigation efforts are warranted. It is primarily through these modeled projections, calculated under various "what if" scenarios, that the risks of anthropogenic global warming begin to take on a recognizable shape. While a historical rise in global temperature of 1°F (.6°C) is certainly significant, it is the alarming future temperature projections generated by the climate models that are likely to drive changes in policy.

Climate change skeptics (as well as a number of climate action advocates) question the reliability of the models. They correctly point out that climate models remain largely unvalidated. It will take decades of observing real world conditions before the projections of the models can be effectively evaluated. As articulated in the influential and generally skeptical book, *Hard Green,* by Peter Huber: "When you are saying something about climate a hundred years hence, the one thing you cannot do is step outside and measure the temperature your model has predicted."[107]

A second criticism of the models is that too much about our climate remains unknown for the projections to be meaningful. No matter how sophisticated the mathematics or how immense the computing power, the quality of any model's output is only as good as the quality of its input data. Right now, the available input data suffers from the uncertainties discussed in prior sections, including the unknown climate effects of cloud formation, solar variation, ocean oscillations such as El Niño, global dimming, increased atmospheric water vapor, the ice-albedo effect, and other positive and negative feedbacks.*

In July 2008, at the request of the Department of Energy, the U.S. Climate Change Science Program issued a 124-page report titled *Climate Models: An Assessment of Strengths and Limitations.* While concluding that climate modeling "contributes to an enhanced understanding of climate relevant processes," the report recognized that "several important aspects of the climate system present especially severe challenges to the goal of simulation."[108] Thus, those who urge caution concerning climate model projections have a valid point. However, as with many other issues within the framework of the larger global warming debate, that legitimate skepticism loses its value when taken to the extreme.† The complete rejection of any potential predictive capacity of the models

---

* Another potential negative feedback involves a "natural heat vent." In a nutshell, it is theorized that warmer temperatures over the so-called warm pool in the tropical Pacific Ocean would limit the formation of heat-trapping cirrus clouds, allowing vast amounts of heat to dissipate into space.

† For example, some version of the following rhetorical question is echoed on several climate-skeptic Web sites: "How can we expect climate models to tell us what is going to happen in fifty years, if they can't predict what's going to happen next week?" The argument belies a fundamental misunderstanding of the difference between weather (which is by nature short term and chaotic) and climate (which represents average weather patterns over longer periods of time).

is no more reasonable than a blind adherence to the models themselves.

For all of these reasons, skepticism should be embraced—not ridiculed. There is much about the climate yet to be revealed, and science unexposed to intense challenge is unreliable. That is not to say, however, that cynicism, ignorance, or inaction should win the day. To the contrary, it is the *risk*, not the *certainty*, of climate change that does or does not warrant action. A homeowner does not buy insurance knowing his house will burn, and humanity need not wait until its fate is sealed before pursuing prudent preventative action. Before evaluating some of the risk mitigation options currently on the table (in Chapter 8), we consider one final, highly controversial component to the sustainability challenge: how population growth will impact GHG emissions.

7

Population and Sustainability
Smaller Footprints/More Feet

The objectives of any global warming mitigation strategy are the stabilization and eventual reduction of greenhouse gases at the top of the troposphere. The level of anthropogenic GHGs discharged during any period is the product of two variables: per capita emissions and the number of human beings on Earth. For example, if, by 2030, per capita emissions are reduced by 25%, but during that same period the world's population increases by 25%, we will not have even *slowed the effects* of man-made climate change.

THE RELATIONSHIP BETWEEN GHG EMISSIONS AND POPULATION

Because per capita emission reductions and population stabilization are two sides of the same sustainability coin, any consideration of one without the other is necessarily incomplete. As suggested by one ecological think tank: "The most effective national and global climate change strategy is limiting the size of the population. Population limitation should therefore be seen as the most cost-effective carbon offsetting strategy available to individuals and nations."[109] While it is generally acknowledged that a growing world population will intensify the challenges of mitigating climate change, many scientists, environmentalists, and politicians are reluctant to openly discuss this highly charged subject. Human population growth and the related hot-button issue of reproductive rights are often referred to as "the elephants in the room" and are considered by many to be "the great taboo of environmentalism." There are signs, however, that as the prospects of global calamity become more imminent, these issues will find their way into the mainstream public debate. Before taking a closer look at the controversial issue of population control, it is worthwhile to consider the astounding growth of mankind over the last ten thousand years.

In 8000 BC, human population stood at 5 million—roughly that of the Metro Atlanta area today.* As our ancestors adapted from hunter-gatherers to farmers, their numbers began to grow at an exponential rate. By AD 1, our population had multiplied sixty times over, reaching 300 million—roughly that of the United States today (Figure 12). That number doubled to 600 million people by AD 1600, doubled again to 1.2 billion by 1850, and doubled yet again to 2.4 billion in 1950. Since 1950, the world's population has grown to 6.8 billion. Each year now produces a net growth of 76 million humans, a number equivalent to the combined populations of Poland and Argentina.

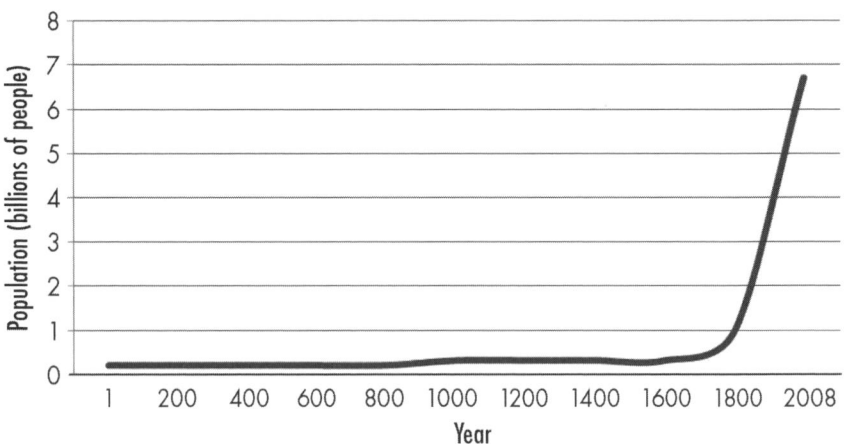

FIGURE 12. World Population Growth *(Source: U.S. Census Bureau)*

---

* All population figures cited here taken from the U.S. Census Bureau website. www.census.gov/ipc/www/worldhis.html

CLARION CALLS

English economist Thomas Malthus is generally credited with establishing the study of the relationship between population and sustainability. Malthus grew up in a privileged home with a utopian father, and, like many sons, rebelled against his father's teachings. In 1798, at age thirty-two, Malthus published his *Essay on the Principle of Population* which criticized the utopian notion that wealth transfer from the rich to the poor could ever end poverty. His central thesis was that, regardless of wealth, unchecked population growth is destined to exceed the means of subsidence, and that only "moral restraint" could avoid the consequences of "misery" such as starvation and disease. Malthus was particularly prescient in his belief that universal suffrage and public education would help alleviate overpopulation—a theory whose application is discussed later in this chapter. Over the last two centuries, the notion that every society (including the world as a whole) has limited resources, and that those limitations present a Malthusian "trap" or "check" on population, has been vigorously debated.

Another prominent but more contemporary voice on the dual issues of population and sustainability is that of Paul Ehrlich, an American professor of Population Studies at Stanford University. In 1968, Ehrlich published *The Population Bomb,* a book that contains a series of controversial, groundbreaking, and, as it turns out, spectacularly wrong predictions about human sustainability. In that book and elsewhere, Ehrlich foretold of an array of dire consequences, including global famine and a reduction of population by 1985 "to some acceptable level, like 1.5 billion people," and boldly stated that he would "take even money that England will not exist in the year 2000." While some credit Ehrlich for igniting debate on an important topic, he is most often cited as an example of the dangers of alarmism and hysteria.[110]

In his acclaimed 2005 best-seller, *Collapse: How Societies Choose to Fail or Succeed,* Jared Diamond, an American evolutionary biologist*

---

* Diamond's fields of study also include environmentalism, geography, anthropology, ornithology, and linguistics.

and National Medal of Science winner, took a different approach, mining history for guidance on future sustainability. *Collapse* tells the stories of several past societies confronted with serious environmental challenges (some that successfully adapted and others that did not) and draws powerful analogies to today's global sustainability issues—including climate change. The societies featured by Diamond include two Pacific islands, Easter Island and Tikopia, which faced similar problems but experienced far different fates.

Easter Island, the world's most remote habitable location, is famous for its enormous stone statues (Moai) which, on average, stand fourteen feet tall and weigh fourteen tons. When this outlying island's long isolation was ended by a Dutch explorer on Easter Sunday 1722, its population had fallen off by 80% and its civilization had drastically degenerated—leading the Dutch discoverers to question how the islanders could have possibly constructed the Moai.

The prevailing view now holds that about a century before the Europeans' arrival, Easter Island experienced a haunting, self-inflicted, ecological disaster. The islanders had apparently felled palm trees to roll the massive statues, build canoes, and clear farmland until the trees were simply gone. This degradation caused a loss of topsoil (and thus crops), wildlife, and canoes for fishing; which, in turn, led to civil war and cannibalism. Thus, the highly advanced culture which had built the Moai between AD 1400 and 1600, it would seem, had effectively committed eco-suicide.*

Tikopia, a tiny Polynesian island occupying a mere two square miles, is home to approximately 1,200 people. To maintain a sustainable food supply amid sparse resources, Tikopians have historically enforced a zero-population growth strategy that included infanticide, suicide, abortion, occasional expulsion (into the sea), and an enforced system in which only first-born sons can marry and have children.[111] While the survival of the Tikopian civilization is ultimately a success story, it provides grim insights into what a true Malthusian check—where finite resources become fully exploited—looks like.

---

*A rebuttal to the view that the Easter Island civilization had collapsed *before* European arrival written by Benny Peiser can be found at www.sacredsites.com/americas/chile/easter_island.html

POPULATION CONTROLS

Certainly Tikopian population control methods are antithetical to our contemporary view of human rights and liberties, but it is not necessary to travel to a tiny island in the middle of the Pacific Ocean to find examples of coercive reproductive restrictions. Since World War II, India has been home to one of the world's fastest growing populations. In April 1976, Indian prime minister Indira Ghandi promoted a policy of compulsory sterilization by vasectomy, reasoning: We must act decisively and bring down the birth rate. . . . Some personal rights have to be kept in abeyance for the human rights of the nation: the right to live, the right to progress."[112] To implement this policy, a variety of coercive sterilization measures were enacted at both state and local government levels. While the laws varied from jurisdiction to jurisdiction, they included: requirements that any man with three children submit to a vasectomy, payment to informants, forced abortion, financial incentives to submit to a "voluntary" vasectomy, and the withdrawal of governmental housing and medical benefits to anyone who did not comply.[113] By early 1977, shortly before Ghandi lost a bid for reelection, national sterilization levels had reached an astounding one million per month, and efforts to meet quotas had led to widespread abuse—particularly victimizing the poor.

While India's forced sterilization program of the 1970s and China's existing one-child policy are authoritarian in nature, not all efforts to turn the "population knob" are coercive. Access to contraception and basic reproductive education are strictly voluntary measures. The United Nations Population Fund estimates that 350 million reproductive-age women lack access to a range of safe and effective contraceptive methods.[114] In sub-Saharan Africa, where much of the world's population growth will occur over the coming decades, 24% of married women have an unmet need for contraception, meaning they are sexually active but do not want children for at least two years.[115]

ALTHOUGH THE WORLD'S POPULATION IS INCREASING, THE RATE OF GROWTH IS DECREASING

While most population statistics are ominous, there are two important figures from which we can take some comfort. First, the average number of children born per female worldwide decreased from 5.02 to 2.65 between 1950 and 2005. The birthrates in Europe (1.41) and North America (1.99) are now low enough that, without immigration, current population levels would be maintained. Second, the world's annual population growth rate peaked in 1989 with a net increase of 87 million people. By 2004, the rate of expansion had been reduced to 76 million per year,[116] and a United Nations study anticipates that global population will level off by 2050 (with a medium-peak population estimate of 9.1 billion).[117]

Various factors are contributing to the decline in birth rates in the industrialized world, including education, the modern trend of women's work outside the home, and a decreased demand for youth labor in urban settings. It is anticipated that, as higher standards of living take hold in the developing world, these influences will lead to a similar stabilization of the population in those nations. The generally accepted sociological concept that industrialization invariably leads to population equilibrium is known as the theory of demographic transition.[118]

Even if the demographic transition process continues as anticipated, the world's population is expected to grow by 2 to 3 billion by 2050. It is universally recognized that this population surge will be a substantial driver of increased GHG emissions. Nevertheless, because of the controversial relationship between birth rates and reproductive rights, climate change strategies rarely even acknowledge population as an unsettled variable. Given the long, sordid history of coerced population control, some trepidation here is understandable. Ultimately, however, it seems curious that some international leaders would deem the threat of global warming so grave as to justify a major realignment of the world economic order, yet not so consequential as to merit open debate on the issue of *voluntary* population controls.

Governments Step In—
Except When They Don't
Solving the "Problem
of the Commons"

Climate change is not a problem that will be solved at a local or national level. Regardless of their point of origin, GHGs mix quickly into the ambient air and spread across the entire atmosphere. At the top of the troposphere where the greenhouse effect occurs, CO_2 levels above a coal-fired power plant are not much different than above virgin forest land. Because of the manner and speed in which GHGs diffuse into the air, any nation or group of nations with sufficient industrial capability has de facto veto power over any proposed international solution. Put simply, if the global community is to reduce emissions to the levels required to avoid dangerous climate change, it will be necessary to engage in a worldwide cooperative effort far beyond anything that human beings have ever before attempted.

THE TRAGEDY OF THE COMMONS

In his classic 1968 essay, *The Tragedy of the Commons*,[119] Garrett Hardin used the parable of individual herdsmen permitting their cattle to overgraze upon a community's commonly owned pasture. The "tragedy" was the creation of a societal dynamic in which it was clearly in the community's *overall* interest to limit grazing to a sustainable level, but it was in each herdsman's *individual* interest to maximize his cattle's use of the common land. Hardin's essay aptly captures the essence of global warming's most imposing problem: Even if all of the world's nations recognize that it is in their collective interest to reduce GHG emissions, it is not in their individual interests to do so unilaterally.

Harding suggested that the tragedy of the commons can be solved in one of two ways. First, commonly owned land could be transferred to private ownership which would incentivize better stewardship. Second, he suggested the implementation of some method of "mutual coercion mutually agreed upon." As applied to climate change, the first approach is untenable because the atmosphere cannot be effectively privatized. Therefore, if the human community is going to achieve the global emission reductions necessary to stabilize the climate, it will require some form of coercive international agreement. This chapter will focus upon what such a worldwide agreement might look like.

AVOIDING DANGEROUS CLIMATE CHANGE

Science cannot identify a precise point at which a rise in average global temperatures will lead to disaster or misery. Several sources, however, including the European Union, have established an increase of 3.6°F (2.0°C) over pre-industrial levels as the threshold—beyond which lies great potential for dangerous climate change.[120] According to a September 2007 report titled *How to Avoid Dangerous Climate Change*, prepared by the Union of Concerned Scientists (UCS):

> Substantial scientific evidence indicates that an increase in global average temperature of two to three degrees Celsius above pre-industrial levels (i.e., those that existed prior to 1860) poses severe risks to natural systems and human health and well-being. Sustained warming of this magnitude could, for example, result in such large-scale, irreversible changes as the extinction of many species and the destabilization and extensive melting of the Greenland and West Antarctic ice sheets—causing global sea level to rise between 12 and 40 feet.[121]

The UCS report also concluded that to "maintain a reasonable chance" of keeping temperatures within this range, atmospheric GHGs must be stabilized at 450 ppm CO_2 equivalent. Stabilization at these levels would provide a medium chance (about 50%) of avoiding the 3.6°F (2°C) threshold increase, beyond which severe consequences are likely to await.

> The global warming potential or GWP of a greenhouse gas is its heat-trapping ability. The most common unit of measurement for GWP is the carbon dioxide equivalent or CO_2 eq. Carbon dioxide is assigned a GWP of 1 and is used as a reference to determine the GWP of other GHGs (e.g., the GWP for methane is 24.5). The GWP impact of all GHGs is measured and reported in terms of CO_2 equivalency. While the concept is straightforward, its application is not. Because certain GHGs are discharged into the air along with aerosols that have a cooling effect, determining the true atmospheric CO_2 equivalent at any point in time presents a challenge.

According to the UCS, in order to achieve the 450 ppm stabilization target, global emissions will need to be reduced "on the order of 40 to 50% below 2000 levels by 2050.[122] While a 40 to 50% reduction is a tall order for a mature economy with access to cutting edge renewable energy, it simply cannot be achieved by a developing nation without: (1) substantial economic and technological assistance or (2) massive regression into poverty. As Lu Xuedu, the deputy director of China's Office of Global Environmental Affairs stated at an October 2006 conference: "You cannot tell people who are struggling to earn enough to eat that they need to reduce their emissions."[123] Thus, a principal challenge in negotiating a worldwide climate agreement will be finding a workable balance between the respective abilities and needs of the developed and developing world.

CAP AND TRADE OR CARBON TAX

All signs now point to a future in which manufacturers and power plants will have to pay for the "privilege" to emit carbon. As discussed in the sections that follow, many nations already participate in cap-and-trade programs designed to reduce emissions, and several other countries are poised to do so. However, while cap-and-trade solutions are clearly the leader out of the gate, an alternative approach to reducing emissions, the carbon tax, has received substantial support, particularly in Australia, Sweden, Ireland, and parts of Canada and the United States.

A cap-and-trade system seeks to mitigate global warming by setting a cap on the amount of CO_2 that can be emitted. A governing body issues or auctions emissions permits equal to the amount of the cap. The permits can then be used by an emitter or traded. A carbon tax is much simpler: A CO_2 emission fee is imposed, usually on a per ton basis, at some point along the fuel supply chain or at the point of combustion.

The most important distinction between a cap-and-trade program and a carbon tax is that each provides different types of certainty. Cap-and-trade ensures a certain outcome regarding how much carbon enters the atmosphere but results in uncertainty regarding the cost to the emitter. The price of emission permits can be volatile, which has the potential to chill investment in renewable projects. Conversely, a carbon tax provides greater price stability but cannot guarantee a precise level of emission reduction.

Proponents of a carbon tax also point out that it is easier to administer; more transparent; and, unlike a cap-and-trade program, easily applied to the transportation sector. Despite these advantages, and despite strong support from economists,[124] the carbon tax strategy suffers from two potentially lethal afflictions. First, it is called a "tax." Although it seems absurd that one of history's most important decisions may ultimately come down to choosing the option with the more pleasant name, that is precisely what many in the press and in government are suggesting.[125] According to the UN's top climate official, Yvo de Boer, those who support a carbon tax are "walking uphill": "If you were a pure economist, the most logical thing is taxation. It is the simplest. But 'taxation' is a word that makes people choke in normal times. And these are not normal times."[126]

Second, the carbon tax proposal simply does not have the political momentum that the cap-and-trade option has accumulated. In the United States, President Obama and Congressional Democrats have cast their lot with the cap-and-trade approach, and, at least for the time being, the proponents of a carbon tax find themselves on the outside looking in.

THE KYOTO PROTOCOL

In June 1992, three years after James Hansen offered landmark Congressional testimony that man-made global warming was almost certainly a reality, leaders of the world convened in Rio de Janeiro, Brazil, to address a host of international environmental issues. That United Nations conference, now commonly known as the Rio Earth Summit, produced the first building blocks for a global agreement on GHG emission reductions.

The Rio Summit yielded two significant achievements. First, it resulted in a treaty highlighting goals for reducing worldwide GHG emissions. Although compliance with the treaty was voluntary, it was viewed by many as an important symbolic gesture. The second, and more remarkable, product of the Rio Summit was the creation of the United Nations Framework Convention on Climate Change (UNFCCC). To date, most international effort to limit atmospheric GHG emissions through "mutual coercion mutually agreed upon" has occurred under the auspices of the UNFCCC.

The UNFCCC roadmap calls for annual conventions, referred to as Conferences of Parties (or COPs). The third such conference (COP-3) took place in Kyoto, Japan, in 1997. The ultimate legacy of the Kyoto Conference was the world's first legally binding international agreement to reduce GHG emissions: the Kyoto Protocol.

In general terms, the Kyoto Protocol requires the world's industrialized nations to reduce their collective greenhouse gas emissions by 5.2% as compared to the baseline year of 1990. Each nation, however, negotiated its own national rate of reduction (or in some instances, its own rate of increase). By way of example, the United States agreed to a 7% reduction, Germany a 21% reduction, Greece a 25% increase, and Russia agreed not to exceed 1990 levels. All of these commitments run until the treaty expires in 2012.

The most controversial aspect of the Kyoto Protocol is the difference in treatment between countries designated as industrialized nations and countries designated as developing nations. The simple truth is that industrialized nations bear all of the sacrifice under the agreement and the developing nations are the beneficiaries of that sacrifice. Although the developing nations participate in the Kyoto Protocol, they do so in ways that are completely voluntary and generally beneficial to their own interests.

The Kyoto Protocol is a cap-and-trade agreement, under which each participating industrialized (Annex I) nation has agreed to a limit on the amount of carbon that it will emit. Within these nations, businesses (sometimes called "operators") obtain, by assignment or auction, allowances to emit certain levels of carbon. If an operator's emissions exceed its assigned allowances, the business has the option of purchasing the unused allowances of another operator or obtaining something known as a carbon credit. While allowances (also known as emission permits) are finite, the number of potential new credits is not. Essentially, under the treaty's credit mechanisms, the creation of sustainable energy projects in participating industrialized or developing nations provide a set-off against emissions.

These credit mechanisms were designed to serve two important functions. First, they encourage market-based efficiency because operators and investors will have more incentive to implement emission reduction projects wherever the best opportunities present themselves. Because a one-ton emission reduction in Bombay is just as significant as a one-ton reduction in Liverpool, the location of the mitigation project is immaterial. Second, because many of these projects will occur in developing nations, those nations will gain access to, and experience with, technological innovations in renewable energy and sustainability.

KYOTO IN ACTION

Even the most ardent supporter of the UNFCCC process must concede that the Kyoto Protocol is a work in process. The United States Senate never ratified the treaty,* and Canada† and Australia‡ have each wavered on their respective commitments. Emerging economic giants China and India, along with 80% of the world's other nations, maintain the right to emit GHGs at will under the protocol. Of those thirty-six nations that have agreed to emission reductions, one third are former Eastern Bloc countries that would have difficulty exceeding their allotted caps even if they tried. And finally, within the group of nations that have assumed an obligation to make meaningful emission cuts, several have failed. Spain, Italy, and Denmark have substantially missed their targets, and in 2008, Greece was suspended from trading carbon credits for violating GHG reporting rules.[127]

To make matters worse, the protocol's carbon credit procedure, referred to as the Clean Development Mechanism (CDM), has come under fire for its inefficiency and susceptibility to abuse. In order for a CDM carbon credit project to qualify, it must provide quantifiable emission reductions beyond that which would have occurred in absence of the project. In application, however, with expanding economies and constantly changing populations, this "quantification" is rarely more than an educated guess. As British

---

* While the United States is a signatory to the Kyoto Protocol, it is not a participant. On July 25, 2007, after the treaty was largely negotiated, the U.S. Senate expressed its dissatisfaction with the proposed agreement by unanimously approving the Byrd-Hagel Resolution. Byrd-Hagel stated, among other things, that it was "the sense of the Senate" that the U.S. should not be a signatory to the protocol. Notwithstanding that resolution, vice president Al Gore signed the Kyoto Protocol on November 12, 1998. The treaty has not been submitted to the Senate by Presidents Clinton, Bush, or Obama.

† In 2006, Canada's environmental minister announced that Canada had no chance of meeting its Kyoto targets and would pursue other, "more realistic" goals for cutting GHGs. "Canada Supports Six-Nation Climate Change Pact: Ambrose," CBCnews.ca, April 25, 2007, www.cbc.ca/canada/story/2006/04/25/ambrose060425.html

‡ Australia opposed the treaty until a change in leadership occurred in December 2007. The new prime minister has stated that Australia is unlikely to meet its commitments and will likely face penalties. "Australia Ratifies Kyoto Global Warming Treaty," MSNBC.com, December 3, 2007, www.msnbc.msn.com/id/22081582/

Petroleum chairman Sir John Browne summarized: "In principle, the CDM was a good idea. In practice, it has become tangled in red tape and has required governments and investors to do the impossible: estimate the level of emissions that would have occurred in the absence of a project and then to calculate the marginal effect of their actions."[128]

Despite its limitations, the Kyoto Protocol does have its staunch defenders. They point out that while the treaty is an important symbolic expression of the world's commitment to climate sustainability, it is also much more. Since the treaty went into effect, most industrialized nations have begun the mitigation process, more than 1,700 carbon reduction projects have been registered in developing countries, and over one billion dollars has been deposited into the UN Adaptation Fund. Even more so, however, advocates of the Kyoto agreement stress that it was always intended to serve as the first step in a long, complex process. As one commentator put it: "From the outset, Kyoto was an interim measure. It was a radical departure from the previous model of inevitable economic growth. It was always recognized that two further phases were needed, the second bringing drastic cuts in emissions and the third involving far more countries."[129]

Whether the legacy of the Kyoto Protocol will be that of a strong foundation upon which future progress was made possible, or that of a costly step in the wrong direction, is currently unknowable. What is clear is that the Kyoto treaty does not represent a long-term solution.

BALI, COPENHAGEN, AND SHARED VISION

In December 2007, the nations of the world met for the 13th COP in Bali, Indonesia (COP 13). After two weeks of intrigue, insults, tears, jeers, and other high drama—including a mass booing of the American delegation, followed by a turnabout in its position—COP 13 produced the Bali Roadmap. Also known as the Bali Action Plan, the Bali Roadmap represents the biggest breakthrough in international climate negotiations since the Kyoto Protocol. The Bali agreement is referred to as a "roadmap," because it charted a two-year course designed to culminate in a new global climate change treaty at the 15th COP in Copenhagen, Denmark (December 2009).

The Bali Action Plan is centered upon four main building blocks: (1) mitigation (reducing GHG emissions or enhancing carbon sinks), (2) adaptation to climate change that the world fails to mitigate, (3) technology transfer from the developed nations to the developing world, and (4) financing and investment assistance flowing from industrialized nations to developing nations.* The conceptual mortar designed to hold these four blocks together is a nebulous ideal known as "shared vision." In the abstract, shared vision represents a collaborative commitment among all nations to find common ground and to develop long-term cooperative action on climate change. In operation, however, post-Bali efforts to convert the narrative understanding on shared vision into some type of concrete numerical agreement have sputtered. Consideration of the respective positions of a few of the key players at COP 14, the halfway point between the Bali and Copenhagen conferences, provides a sobering glimpse of the work that remains:

- Japan and many industrialized nations believe that all parties should adopt the long term goal of achieving at least a 50% reduction of greenhouse gases by 2050. While industrialized nations should lead the way in reducing net emissions, "economically advanced" developing nations that are large emitters (like China, India, and Brazil) must also accept binding targets.[130]

- China and India do not believe any form of target reduction for developing nations is appropriate at this point. They assert that it is the industrialized nations that bear "historic responsibility" for today's atmospheric GHG levels, and they should unilaterally engage in deep emission cuts. For now, developing nations

---

* The Bali delegates also reached an agreement to include a controversial program known as REDD (Reducing Emissions from Deforestation and forest Degradation) in the negotiations leading up to Copenhagen. Under the REDD program, emitters would pay developing nations to preserve forests. The underlying rationale for REDD is that the preservation of forests is an inexpensive way to sequester carbon.

should participate by planning for the sustainable development that will be supported by finance and technology grants from developed countries. In the end, stabilization must be achieved on the basis that "each human being has an equal right to the common atmospheric resource" (*i.e.,* emissions limitations should be calculated on a per capita basis, rather than on a historical or GDP basis).[131]

As the immense challenges presented by these deep philosophical and economic divides have taken center stage, the optimism inspired by the eleventh hour success at Bali has faded. In June 2009, the UN's top climate official, Yvo de Boer, expressed publicly what many already knew: Progress toward a comprehensive Copenhagen deal was not on track. While remaining optimistic that a solid foundation could come out of Copenhagen, de Boer acknowledged: "I don't think in Copenhagen we're going to get an agreement on an 80% global emission requirement [by 2050] and I think that, at the end of the day, that is what we need."[132]

At COP15 in Copenhagen (December 7–18, 2009), two weeks of raucous negotiation yielded sparing progress. Although the "Copenhagen Accord" represented a few small steps forward, it includes no binding emission limitations. These negotiations made clear that, as the world moves towards COP16 in Mexico, two interrelated questions will dominate the debate. First, will the U.S. participate in a global climate treaty that is disadvantageous to its short- and mid-term economic interests? Second, is China prepared to enter into a global climate treaty that will require it to throttle back its rapid economic growth? If the answer to either of these questions is "no," the prospects of a comprehensive international agreement are negligible.

THE ELEPHANTS IN THE ROOM—THE UNITED STATES AND CHINA

Any international climate deal that is built upon the four pillars of the Bali Action Plan—mitigation, adaptation, technology transfer, and finance—will have the practical effect of transferring jobs and/ or resources from developed nations to the developing world. Especially in the early years, most emissions reductions (mitigation) will be accomplished by industrialized nations. And, the sustainable development and adaptation that does occur in developing nations will be supported by assistance from the developed world in the form of financing, technology sharing, and other forms of capacity building. As a leading South African environmental official candidly stated in August 2009: "No money, no [climate] deal. . . . We need the support, the financial and technological support."[133]

As a result, a successful global agreement would require countries like the U.S. to incur higher relative energy costs, fund adaptation projects in poorer nations, share its proprietary sustainable technology, and finance energy infrastructure in nations such as China and India. Putting aside the question of whether this approach is justified by the "historical responsibility" of the industrialized nations or the "immutable right of developing nations to eradicate poverty," there is little reason to believe that this is a politically feasible solution for the United States.

Article II, section 2, of the U.S. Constitution states that the president shall have the power to make treaties "provided two thirds of the Senators present concur." Given the present political climate, obtaining such a super-majority seems highly unlikely. As climate change author and former U.S. Energy Department undersecretary Joseph Romm wrote: "[E]ven if there are 60 Senate votes to override a right-wing filibuster against a strong domestic climate bill, there aren't 67 votes for a new climate treaty. And that means the UNFCCC process as we now know it is essentially a Dead Man Walking, even if nobody knows it yet."[134]

While circumstances can change at the drop of an ice shelf, persuading sixty-seven Senators to concur on a climate agreement that would ship jobs, technology, and investment to China and India is hard to fathom.

If the UNFCCC process is in trouble, where does that leave us? Outside of the United Nations negotiations, there are at least two viable options. The first is to continue to foster a series of unilateral, national carbon reduction programs with the hope that they can be integrated internationally at some future date. The European Union has already committed to reduce carbon emissions by 20% off of 1990 levels by 2020, and Japan has agreed to a target reduction of 15% by 2020 using 2005 as a baseline.[135] Canada (20% below 1990 levels by 2020) and Australia (5 to 15% below 1990 levels by 2020) have also adopted targets.[136]

On June 28, 2009, by a vote of 219 to 212, the U.S. House of Representatives passed a cap-and-trade bill known as the American Clean Air and Security Act (ACES).* If the Senate passes some version of the ACES bill and it is signed by the president, it would lower U.S. GHG emissions by 17% over 2005 levels by 2020, and by 83% over those levels by 2050. If and when, through independent or concerted action, the number of nations taking concrete steps to reduce GHGs reaches a critical mass, options like political pressure, import tariffs, and international trade agreements will be available to influence remaining holdouts. By way of historical analogy, on November 4, 2004, Russia belatedly ratified the Kyoto Protocol in exchange for the European Union's support of its bid for admission into the World Trade Association.[137]

Should the UNFCCC process stall, the second viable option is bilateral or multilateral talks involving either "the big two" or other key constituencies. The two largest CO_2 emitters on the planet are China and the United States, each accounting for about 20% of global emissions. No other nation emits even one third of that amount.[138] In many respects, these two eight-hundred-pound gorillas represent opposite ends of the climate negotiations spectrum. The United States is a modernized society with very high per capita emissions† and a moderate but stable rate of economic growth. China, although certainly an economic powerhouse, is still

---

* This bill is alternatively known as the American Clean Energy and Security Act, the Waxman-Markey bill, and H.R. 2454.

† The U.S. per capita emission rate is second only to Australia, and more than double that of most European nations.

a developing nation. Despite the fact that China's gross domestic product (GDP) ranks second in the world, its GDP per person (of about $3,000)[139] ranks 104th, well below the world's average. For the last quarter of a century, China has maintained a torrid annual average economic growth rate of 9% per year. To maintain its growth, and to lift its population from poverty, China will need to acquire and consume ever-increasing supplies of energy over the coming decades.

If China and the U.S.—large emitting nations on different sides of the debate—could broker a joint GHG reduction plan, it is entirely possible that the rest of the pieces would fall into place. Given China's economic needs, and given its understandable insistence that each Chinese citizen has the right to emit as many GHGs as U.S. citizens, such talks would be, in the words of climate commentator Joe Romm, "the most difficult and most important negotiations in U.S. and world history."[140] An agreement between these two nations would require that either: (1) the U.S. accept conditions that would put it at a substantial competitive disadvantage as to emerging economic powers, during a period in which the U.S. is already hemorrhaging jobs to those countries; or (2) require China to put the brakes on the hyper-growth that has been the centerpiece of its successful economic strategy for decades. The hard truth is that neither nation has yet demonstrated a conviction that the risks of global warming warrant the economic sacrifice that a climate agreement with the other would require.*

Whether it be some future United Nations COP, a G8+5 meeting, the Asian Conference, or bilateral discussions between the U.S. and China, humanity is sailing into uncharted territory. Will the nations of the world find enough common ground to impose "mutual coercion mutually agreed upon" in time to avoid the "tragedy of the

---

*On July 28, 2009, after two days of high-level talks between the United States and China on economic and energy issues, these two nations executed a "memorandum of understanding" on "Climate Change, Energy and the Environment." Although the memorandum reflects no substantive breakthroughs, it calls for "regular consultations" and "cooperation" on climate change and renewable energy issues. www.state.gov/r/pa/prs/ps/2009/july/126592.htm

The "Group of Eight" or G8 is a forum made up of eight nations of the Northern Hemisphere: Canada, France, Germany, Italy, Japan, Russia, the United Kingdom, and the United States. On several occasions, G8 meetings have included additional nations referred to as the Outreach Five (O5) or the Plus Five: Brazil, China, India, Mexico, and South Africa. Since the G8-O5 nations account for a very large percentage of global emissions, this forum represents an alternative venue in which climate change issues may be addressed.

commons"? The answer to that monumentally important question will turn, in large part, upon how quickly cost-effective renewable energy sources can replace fossil fuels. The transition—from the three hydrocarbons to the "sustainable smorgasbord"—is discussed in the pages that follow.

9

Peak Oil
Global Warming's Evil Twin

No analysis of climate change is complete without consideration of how the accelerating depletion of world oil supplies is likely to impact CO_2 emissions and further complicate already tumultuous times. On the one hand, there is a substantial overlap between the sustainability objectives of those promoting energy security and the goals of those seeking to mitigate climate change. Great synergies exist between these two growing constituencies because the development of clean energy solutions serves the dual benefit of reducing GHGs and conserving dwindling fossil fuels. On the other hand, if the world's oil demand outpaces the available supply before a sufficient transition to alternative energies has occurred, nations will inevitably turn to high-carbon unconventional options such as heavy crude, tar sands, oil shale, and coal liquefaction. Large-scale exploitation of these carbon-heavy fuels would likely increase atmospheric GHGs even more than in the IPCC's "high-end" emission scenario, and thereby substantially increase the possibility of climate catastrophe. Thus, while efforts to avoid global warming and efforts to slow oil depletion may share many common solutions, a scarcity of oil before ample renewable solutions are in place would put humanity between a rock and a hard place.

OIL—THAT IS, BLACK GOLD, TEXAS TEA

Oil is the lifeblood of modern civilization. It fuels our air, seas, and ground transportation. It is an integral component of plastics, cosmetics, pharmaceuticals, and other consumer products. It is also required to produce the fertilizers and pesticides that are essential to today's advanced farming. Over the last century, the abundance of low-cost oil has been a great boon to humanity. As author James Kunstler explained:

> Oil is an amazing substance. It stores a tremendous amount of energy per weight and volume. It is easy to transport. It stores easily at regular air temperature in unpressurized metal tanks, and it can sit there indefinitely without degrading. You can pump it through a pipe, you can send it all over the world in ships, you can haul it around in trains, cars, and trucks, you can even fly it in tanker planes and refuel other airplanes in flight. It is flammable but has proven to be safe to handle with a modest amount of care by people with double-digit IQs. It can be refined by straightforward distillation into many grades of fuel—gasoline, diesel, kerosene, aviation fuel, heating oil— and into innumerable useful products: plastics, paints, pharmaceuticals, fabrics, lubricants.[141]

Despite its many virtues, oil suffers from one major drawback: The Earth's endowment of oil is finite. Oil, like other fossil fuels, was formed over millions of years by the decomposition of dead plants and animals. Since the mid-1980s, new oil discoveries have substantially lagged behind oil production (Figure 13), and it is commonly recognized among geologists that the ever-dwindling supply will not meet the ever-growing demand for too much longer. The fast-approaching moment in time when oil production can no longer be increased to meet the word's increasingly voracious appetite, has been given the name "peak oil."

THE TIMING OF PEAK OIL

In 1956, a geologist named M. King Hubbert introduced a model that accurately predicted that U.S. oil production would peak around 1970 and decline shortly thereafter.[142] Hubbert's premise

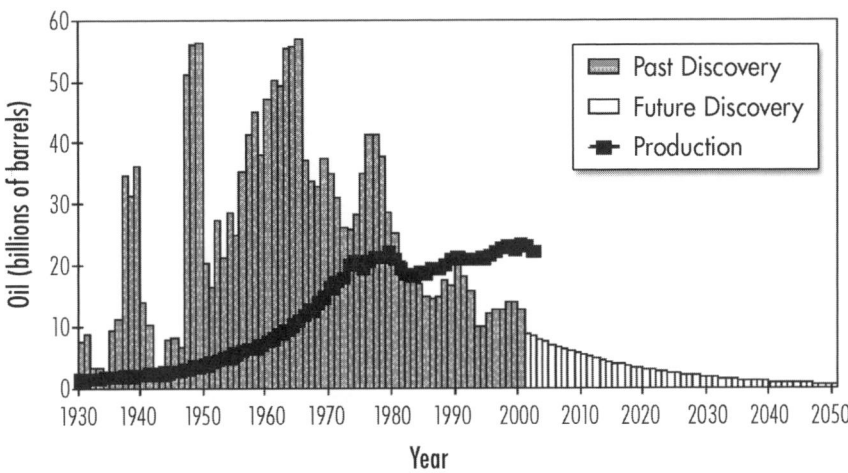

FIGURE 13. The Gap Between Global Oil Discovery and Production

was that the production of oil (and other fossil fuels) in a particular region would follow a bell-shaped curve. After a period of discovery, oil production rates would grow exponentially until half of its reserves were exhausted. At that halfway point, production would peak, plateau, and then decline. The basic premise of Hubbert's theory is now widely accepted, and its bell-shaped curve projection has been used successfully to predict the production peaks and declines of regional fuel and global mineral supplies.[143] The bell-shaped pattern of oil production in the lower forty-eight states of the U.S. is depicted in Figure 14.

There is a general consensus among experts that, in accordance with Hubbert's theory, world oil output will peak at roughly the same time that half the world's recoverable oil has been produced. Great uncertainties exist, however, concerning the extent of undiscovered oil, the amount of oil remaining in known locations, and how new extraction technologies may increase global supply. Because of these complicating factors, the precise timing of peak oil won't be known for sure until the world sees it in its rearview mirror.

Several respected sources predict that peak oil production is occurring now, or will occur in the very near future. In October

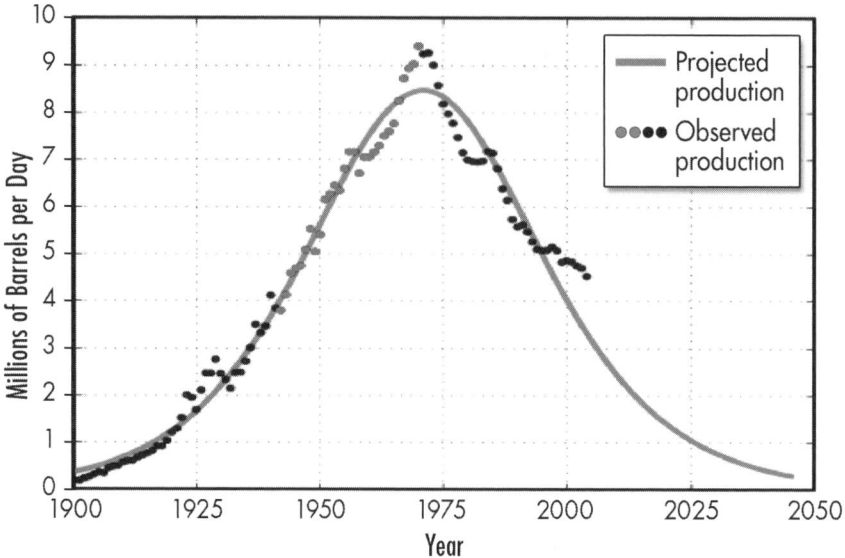

FIGURE 14. U.S. Oil Production, Lower 48 States *(Source: Energy Information Administration [EIA])*

2007, Sadad al-Huseini, former head of production at the world's largest oil corporation (Saudi Aramco), asserted that production had reached a "maximum sustainable plateau."[144] In January 2008, the CEO of Shell Energy Jeroen van der Veer predicted that "after 2015, supplies of easy-to-access oil and gas will no longer keep up with demand."[145] And, in May 2009, U.S. financial services company Raymond James joined the growing list of experts concluding that peak oil had already occurred, declaring that in the first quarter of 2008, oil production "crossed over to the downward sloping side of Hubbert's Peak."[146]

Many other assessments, however, remain more optimistic, suggesting that peak oil is still at least a decade or more away. The Energy Information Administration (EIA) within the U.S. Department of Energy, in its 2009 *International Energy Outlook,* projected a substantial increase in oil production over the next two decades. In that report, the EIA predicted that 2007 global oil production levels of around 81.5 million barrels per day would increase to 93.1

million barrels per day by 2030. In its own understated way, however, the EIA report is very troubling; the 93.1 barrels per day projection is down substantially from earlier EIA predictions made in 2005 (122.2 million barrels per day by 2025) and in 2008 (107.2 million barrels per day by 2030).[147]

THE EFFECTS OF PEAK OIL

While disagreement remains over when peak oil will arrive, very few experts question that the downside of Hubbert's curve is fraught with extraordinary danger. A stagnation and decline of oil supply, especially at a time when developing nations such as China and India are boosting global demand, would almost certainly lead to economic turbulence and new levels of international competition—leading to some combination of economic shock, political disruption, starvation, political unrest, or even war. Over the last century, global economic development has been shaped largely by access to abundant, affordable oil, and absent immediately available alternative energies, it is unclear how advanced human societies will sustain themselves.

A February 2005 report sponsored by the U.S. Department of Energy addressed the potential impacts of peak oil. This document, commonly referred to as the Hirsch Report, after its project leader Robert L. Hirsch, warned of the potential for some very dire consequences. Some of the Hirsh Report's key conclusions are:

- While it is unclear when peak oil will occur, "geologists have no doubt that it will happen." Nine of twelve predictions cited by the Report's authors project the occurrence of peak oil before 2020.

- "Previous energy transitions (wood to coal and coal to oil) were gradual and evolutionary; oil peaking will be abrupt and revolutionary."

- "Oil peaking will create a severe liquid fuels problem for the transportation sector, not [merely] an 'energy crisis'. . . . [and] will cause protracted economic hardship in the United States and the world."

The report concludes that, while the unique challenges of peak oil are not insurmountable, the solutions will require intense governmental effort and the expenditure of trillions of dollars *before* peaking occurs.

In January 2008, Shell CEO Jeroen van der Veer, wrote an open letter to all company employees.[148] In his letter, van der Veer predicted how, under two alternative scenarios, peak oil might impact GHG emissions and climate change. Under the preferred "blueprints scenario," international cooperation on renewable energy solutions, conservation, and carbon sequestration (discussed in Chapter 11) would result in a relatively soft landing. Conversely, under the "scramble scenario," governments choose politically expedient and local solutions. As oil becomes increasingly scarce and difficult to produce, an unprepared world turns to coal and oil sands. The combination of extreme energy costs and dependence upon these high-carbon fuel options throws the planet into decades of economic and climactic volatility. Van der Veer's conclusions echo those of the Hirsch Report: "The sobering reality is that the Blueprints scenario will only come to pass if policymakers . . . promote energy efficiency and technology in four sectors: heat and power generation, industry, mobility and buildings. It will be hard work and there is little time."

In a perfect world, the transition to clean renewable energy would be substantially complete before oil production begins its the inexorable slide down the back half of Hubbert's curve. However, in the world in which we live, it is becoming increasingly likely that peak oil will arrive before the transportation sector has converted to clean electricity, biofuels, or hydrogen (Chapter 12). Although this production gap is expected to trigger greater reliance by the transportation sector on carbon-intense heavy crude, tar sands, and oil shale, another energy option has recently been pushed to the forefront, courtesy of a fiery billionaire from rural Oklahoma.

NATURAL GAS AND THE PICKENS PLAN

Natural gas is the cleanest of the fossil fuels, the so-called prince of the hydrocarbons. It is composed primarily of methane, but also contains ethane, butane, and propane. Per unit of energy, natural gas emits 43% less carbon than coal, and 30% less carbon than oil.[149] As a result, conversion from coal-fired power to natural gas would significantly reduce GHG emissions. In its "dash for gas" in the 1990s, England was able to handily achieve its emission reduction targets under the Kyoto Protocol by replacing many of its coal plants with methane gas facilities.

Natural gas currently produces about 18% of U.S. electricity and about 20% of all electricity worldwide.[150] As recently as 2008, it was widely believed that North American methane production had peaked. According to a June 2009 report by the highly respected Potential Gas Committee, however, natural gas reserves in the U.S. surged an astounding 39% over 2006 levels.[151] Much of this increase can be attributed to innovative drilling technologies that permit recovery from shale rocks. With these newfound increases, U.S. reserves now equate to a one-hundred-year supply of natural gas at current consumption levels (or a fifty-year supply, if consumption were to double).[152]

Despite this very good news, natural gas is not a long-term solution to existing climate change or U.S. energy security challenges. The combustion of methane still emits large quantities of carbon, and its supply is finite. Because natural gas emits less CO_2 than coal and oil, however, and because it is relatively plentiful on the North American continent, it certainly has the capacity to help address these challenges while more-sustainable options are developed over the next few decades.

While it is generally acknowledged that natural gas can serve as a bridge to a renewable energy future, heated debate exists over how it can be best utilized. The simplest approach is to use methane to displace coal-burning power plants. As Joseph Romm explained in June 2009, "Other than energy efficiency and conservation, the lowest-cost option for achieving large-scale CO_2 reductions by 2020 is simply replacing electricity produced by burning coal with power generated by burning more natural gas in the vast array of currently underutilized gas-fired plants."[153]

An alternative approach involves using natural gas as a transportation fuel to replace oil. Compressed natural gas (CNG) has been used successfully as an automobile fuel for over sixty years.[154] There are now approximately 8 million CNG vehicles on the roads around the world, about half of which operate in South America.[155] While the U.S. transportation sector has been slow to embrace CNG, encouragement to do so is now resonating from an unlikely source. T. Boone Pickens, a former Oklahoma oilman, armed with a $58 million advertising budget, has presented a bold series of proposed solutions to America's energy related problems that he has humbly named the "Pickens Plan."

The Pickens Plan differs from most other renewable energy proposals in two critical respects. First, it is an actual plan—not merely a statement of challenges faced or a series of vague goals. Second, while the plan, if implemented, would result in a substantial reduction of GHG emissions, it is first and foremost an energy security strategy.

There are four core components or "pillars" to the plan:

1. Dotting the wind corridor of the American Great Plains with over 100,000 wind turbines capable of generating 20% of U.S. electricity demand.

2. Building thousands of miles of power transmission lines to connect remote wind turbine and solar facilities to those power grids servicing population centers.

3. Providing robust incentives to the owners of homes and commercial buildings to invest in a variety of energy efficiency strategies.

4. Taking the natural gas that has been hypothetically freed up by the increased wind capacity and using it to fuel transportation—particularly large truck and bus fleets.

These steps, according to Pickens, would cut U.S. oil imports by one third, create hundreds of thousands of jobs, and substantially reduce GHGs. While the first three pillars of the plan are meant to be permanent solutions, the natural gas component is conceived as a stopgap measure until hydrogen or electric-powered vehicles become practical (see Chapter 12).

Several aspects of the Pickens Plan—an urgent commitment to wind production, massive investment into transmission lines and grid technology, and aggressive conservation—do not stray far from the beaten path. The plan's innovation lies in fueling a substantial portion of the U.S. transportation sector with natural gas. The critical question is: Assuming the U.S. can increase wind capacity to provide 20% of its electricity demand (no small feat), should that new capacity be used to replace imported oil or should that new capacity be used to replace domestic coal?

Pickens's strategy to replace oil with wind (using natural gas as a surrogate) serves U.S. energy security interests while simultaneously reducing GHGs. It does not, however, reduce GHG emissions as much as replacing coal with wind would. Moreover, the conversion of a substantial portion of the U.S. transportation fleet to natural gas, and constructing the necessary fueling infrastructure, would take more time and money than simply revving up existing natural gas-fired power facilities. So the issue comes down to this: How do we prioritize the dual threats of global warming and peak oil? To those who view the latter as particularly perilous, but would also like to take meaningful steps to protect against the former, the Pickens Plan has much to offer.* As presented in the remaining pages, there are yet other factors to consider before we have to make a choice.

---

* Several commentators have questioned Pickens's motives. After all, his wind, water, and natural gas interests all stand to benefit from the plan, but his substantial investments can also be reasonably perceived as evidence that he is personally committed and has "put his money where his mouth is." No one, except Pickens himself, will know his motivations for sure, but it seems overly cynical to conclude that his aims are not, at least in part, altruistic. In the end, these criticisms are largely mere distractions since the plan should either stand or fall on its own merits.

10

Glimpsing a Sustainable Future
Terawatts, Electric Vehicles, and the Smart Grid

It is difficult to overstate the enormity of the dilemma that now confronts humanity. Population levels continue to swell, developing nations are acquiring increasingly voracious energy appetites, and global reserves of the two relatively cleaner fossil fuels (oil and gas) are in decline. Without the development of other viable energy choices, these circumstances will inevitably lead to one of two extremely unpleasant conclusions.

Behind door number one lays the option of using coal and heavy crude to replace depleted oil and natural gas, dramatically increasing the likelihood that the Earth will experience the devastating effects of rampant global warming discussed in Chapter 5. Behind door number two is the pandemic starvation, civil unrest, and untold misery that would accompany a massive, global energy shortage. While Chapters 11, 12, and 13 discuss particular strategies designed to hastily construct a door number three, this chapter takes a look at several more general sustainability concepts. We start with a primer on how energy is measured, step back to take in the big-picture view of the world's current energy portfolio, consider some of the intriguing possibilities provided by electric vehicles, and evaluate how smart grid technologies will quickly and radically change the way we consume, transport, and store energy.

MEASURING ENERGY AND POWER: ENERGY 101

Energy is measured in joules and is defined as "the ability to do work." A joule is the amount of energy necessary to lift an object that weighs one Newton (about the same as an apple or a stick of butter) one meter off of a surface where it rests. Power is measured in watts and is defined as the rate at which energy is used. A watt is a measurement of the rate (one joule per second) at which energy is converted into power. Here, for purposes of achieving an apples-to-apples comparison between various sources of energy, both production and consumption will be characterized in terms of the more familiar measurement standard of watts.*

The current global demand for power is 15 terawatts. One terawatt is equal to one thousand gigawatts; one gigawatt is one thousand megawatts; one megawatt is one thousand kilowatts; and one kilowatt is one thousand watts (Figure 15). That means, at any given moment, human beings are using 15 trillion watts of power. With a global population of 6.7 billion, the average per person energy use at any instant in time is about 2200 watts (or the equivalent of fifty-four 40-watt bulbs). The average American's rate of consumption is 12,000 watts (or the same as three hundred 40-watt bulbs).[156]

Electric power plants are generally characterized by their maximum output in megawatts (also known as their nameplate capacity). The world's twenty largest power plants each have maximum outputs in excess of 5000 megawatts (5 gigawatts) and are each capable of supplying electricity to several million homes. An equally important, but often overlooked aspect of a power plant's productivity is its capacity factor, which indicates how often the plant is operational.[†] While the capacity factors of U.S. nuclear power

---

* Watts are often confused with watt-hours. Watts are measurements of consumption or production at a given moment. A watt-hour represents that level of consumption or production over a full one-hour period. Thus, turning on a 100-watt light bulb for three hours utilizes 300 watt-hours of power.

† Technically speaking, a capacity factor is the ratio of actual production over time to output if the plant had operated at maximum output.

FIGURE 15. SAMPLE MEASURES OF POWER OUTPUT AND CONSUMPTION		
TERAWATT (TW) (1,000 GW)	**Measures of Global Power Consumption**	
	15 TW	Average current global demand for power (at any given moment)
	1 TW	Average global energy generated by nuclear power (at any given moment)
GIGAWATT (GW) (1,000 MW)	**Measures of Power Output of Large Power Plants and Power Grids**	
	18 GW	Generating capacity of Three Gorges Dam hydro electric plant in China
	5 GW	Electric power needed to heat all the buildings in Maine
MEGAWATT (MW) (1,000 KW)	**Measures of Power Output of Electric Generators and Nuclear Plants**	
	140 MW	Average power consumption of a Boeing 747 passenger aircraft
	7 MW	Generating capacity of world's largest wind turbine
KILOWATT (KW) (1,000 W)	**Measures of Power Consumption of Tools and Machines**	
	12 KW	Average American person's power consumption (at any given moment)
	5 KW	Generating power of typical single home solar system
WATT (W)	**Basic Measure of the Rate at Which Energy Is Converted Into Power**	
	150 W	Average power consumption of a personal computer
	3 W	Power consumption of micro flashlight

plants and coal-fired facilities are quite high (91 and 70%, respectively), wind and photovoltaic solar plants (30 and 20%, respectively) operate much more sporadically.[157] The key point is that nameplate capacity, which is usually the measure set forth in press releases, tells only part of the story. As a result of their different capacity factors, one megawatt of coal fired electricity is sufficient to power six hundred U.S. homes, while a one-megawatt wind turbine may only be in operation enough of the time to supply power to 250 homes.

OUR CURRENT ENERGY MIX, AND LIKELY FUTURE DEMAND

Of the 15 terawatts of energy consumed globally at an average moment in time, the vast majority—13 terawatts—is supplied by fossil fuels (oil, coal, and natural gas). (Figure 16.) Most of the remainder of the world's energy comes from nuclear and hydro-electric facilities, and just over 1% of the world's energy now comes from a combination of solar, hydrogen, wind, and geothermal sources.

The challenge we face is not merely replacing those 13 terawatts of fossil fuel energy with sustainable alternatives. Even with antic-ipated improvements in energy efficiency, overall energy con-sumption will substantially increase in the coming decades due to global population growth and growing demand by developing nations such as China and India (as their living standards contin-ue to improve). Annual world energy consumption has never declined in modern history and is not projected to decline anytime soon. The formidable nature of the task at hand was summed up by *Newsweek* senior editor Sharon Begley in a thought-provoking 2009 editorial titled "We Can't Get There From Here"[158]:

Assuming minimal population growth (to 9 billion people), slow economic growth (1.6% a year, practically recession level)

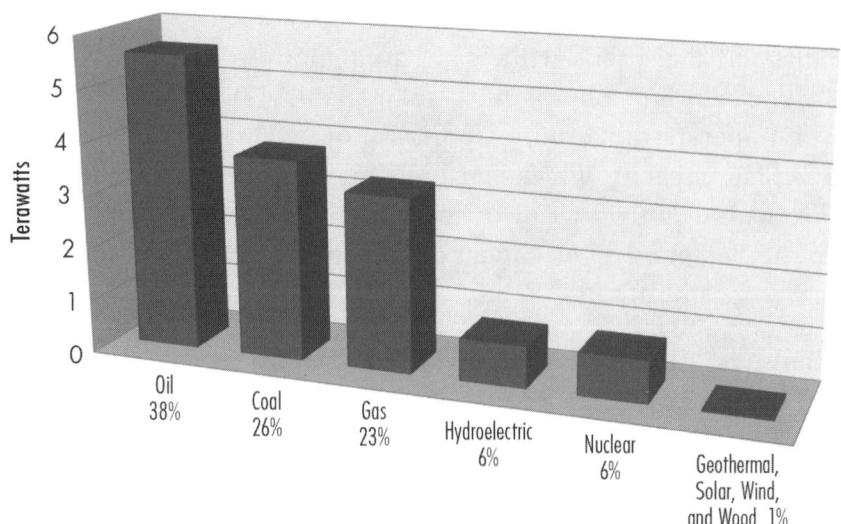

FIGURE 16. World Energy Supply (in Terawatts) *(Source: U.S. Department of Energy)*

and—this is key—unprecedented energy efficiency (improvements of 500% relative to current U.S. levels, worldwide), [the world] will use 28 terawatts in 2050. . . . Simple physics shows that in order to keep CO_2 to 450 ppm, 26.5 of those terawatts must be zero-carbon. That's a lot of solar, wind, hydro, biofuels, and nuclear, especially since renewables kicked in a measly 0.2 terawatts in 2006 and nuclear provided 0.9 terawatts.

Perhaps the single most vexing aspect of what Begley refers to as the "enormous challenge we face" is the transportation sector's complete reliance upon liquid fossil fuels. Part of the solution to that problem may be found in a technology once given up for dead: the electric car.

PLUG-IN AUTOS

Transportation now accounts for 20% of global energy consumption. As access to automobiles grows in developing nations, that percentage is expected to sharply increase. In January 2009, for the first time, more cars were sold in China (735,000) than in the United States (656,976).[159] Given these circumstances, any realistic strategy to reduce GHG emissions will require a revamped automobile design that is less reliant upon oil. Biofuels can be generated to shoulder some of the load, but (as discussed in Chapter 13) their role in reducing GHG emissions will be limited. Hydrogen may be the fuel of the long-term future, but the technical and economic challenges to developing an integrated hydrogen delivery system may not be overcome for decades (also Chapter 13).

The substitute to fossil fuel that now shows the most immediate promise is "plug-in" electric and dual fuel (i.e., gasoline and electric) technology. As sustainable technologies such as wind, geothermal, solar, clean coal, nuclear, tidal, wave, and hydropower contribute more and more energy to the power grid, electric cars will become an increasingly larger part of the solution. Just as early DVD players were sold in combo units that could also play VHS tapes, the dual-fuel feature of these vehicles is designed to facilitate a two-step transition from today's combustion engine to tomorrow's electric motor.

Besides the ability to operate on energy generated by wind, or the sun, or any other renewable source that can produce electricity, these vehicles will be capable of storing unused energy and feeding it back into the grid. This storage capability will make electric cars an integral part of a revolutionary new approach to how energy is distributed, stored, and consumed. This imminent, radical transformation, involving the implementation of a series of breakthroughs referred to as "smart grid technologies," is discussed in the next section.

SMART GRID TECHNOLOGIES

While the term "smart grid" inspires visions of some monolithic control center buried deep within a mountain bunker, the smart grid concept is really just the integration of several technologic and infrastructure improvements. As defined in a December 2008 report commissioned by the U.S. Department of Energy, a smart grid is "a more intelligent, resilient, reliable, self-balancing, and interactive [electric power] network that enables enhanced economic growth, environmental stewardship, operational efficiencies, energy security, and consumer choice."[160]

Before outlining the three principal innovations of the smart grid, a few words about the existing "crazy quilt" system are warranted. Today's electricity distribution systems are antiquated, highly inefficient, often incapable of supporting renewable technologies, and increasingly vulnerable to massive failure or terrorist attack. The root cause of all of these shortcomings is that current grid systems were designed and constructed to facilitate an energy-delivery paradigm that simply no longer exists. For over a century, electric grid design was predicated upon three bedrock assumptions: (1) that power generation facilities would be large and local; (2) that power supply would always flow in same direction, toward the end user; and (3) that consumers would be charged a flat rate per kilowatt hour regardless of the time of day when they used the power. Now that these assumptions are no longer valid, we are going to need a new grid.

The first, and most urgently required, feature of the new smart grid system is the construction of extra high voltage transmission lines to the remote areas where renewable infrastructure is now being built (or would be built if the lines existed). Unlike fossil fuel plants that can be conveniently situated near the cities and towns they supply, the most suitable locations for wind turbines and large solar facilities are open plains and sweltering deserts. Right now, there is simply no way to move that clean renewable energy to the population centers where the demand for it exists—and, too often, new wind and solar capacity is wasted because existing transmission lines are congested.[161] While building new transmission lines will ultimately result in a savings—both in terms of cost and GHG reduction—the required investment will be substantial. In the United States alone, the new infrastructure costs are projected to exceed 100 billion dollars* over the next fifteen years.

The second attribute of new smart grid technologies is that they will facilitate a fundamentally new approach to electricity production, known alternatively as "decentralized power," "distributed generation," or the "democratization of energy." Whatever it is called, the basic idea is that electricity distribution channels will cease being one-way streets. Any home or business with roof-top solar panels, a micro turbine, or any other source of generated or stored electricity will be able to feed power back into the grid. For example, on a sweltering summer day when electricity demand spikes, tens of thousands of homes with photovoltaic solar installations could sell their produced electricity back into the system—reducing strain on the grid.

---

* A February 2009 study performed by regional grid operators, known as the Joint Coordinated Systems Plan, estimated that it would cost in excess of $80 billion in new transmission infrastructure to provide the eastern United States with wind power sufficient to satisfy 20% of its electricity demand by 2024. Joint Coordinated System Plan Press Release, February 9, 2009, http://www.midwestiso.org/publish/Document/20b78d_11ef44fc9c0_-7c4b0a48324a/20090209%20JCSP%20Study%20Quantifies%20Cost%20of%20Delivering%20Wind%20NR.pdf?action=download&_property=Attachment

This two-way transmission and distribution feature will also advance the development of electric vehicles. In the not-too-distant future, battery-powered cars and hybrids will operate as energy storage devices. These vehicles can be charged at night when (as discussed later) both demand and price are lower but wind continues to whip across the prairie, spinning turbines and creating excess energy supply. Then, during the day, when demand and price are higher, some of the electricity stored in those batteries can be used to directly power homes or businesses, or the excess power can be fed (potentially at a profit) back into the grid. As Robert Howard, a vice president at Northern California's Pacific Gas and Electric (PG&E) put it, with this vehicle-to-grid (V2G) technology, "we can merge transportation with the utility network and fundamentally change the way we live and work."[162]

The third innovative feature of the smart grid concept is real-time pricing. Historically, power companies have provided electricity to consumers at a constant, blended price irrespective of market conditions and their own fluctuating variable costs. Soon, however, consumers will be provided with in-home smart meter devices that identify minute-to-minute time-of-use costs. The expectation is that these variable rates will encourage customers to conserve during peak hours, thereby taking substantial strain off of the grid when it is at or near full capacity.

Smart meters are also capable of working in tandem with "smart appliances," "smart sockets," and "smart thermostats." These accessories complement smart meters by automatically adjusting usage based on programmed consumer preferences as costs fluctuate. Thus, a "smart dishwasher" can be programmed to operate at the point in the evening when energy costs dip to their nightly low. While such smart equipment is the wave of the future, the future has already arrived in places such as Austin, Texas; Boulder, Colorado; and parts of California. In each of these locations, smart meter programs are already in place, and pilot programs are planned or underway in dozens of other communities.

The development and implementation of new smart grid technologies and infrastructure are often analogized to the construction of the existing interstate highway system. Before the Federal Highway Aid Act of 1956, national travel was limited to a patchwork of state roads and country thoroughfares. Just as the interstate highway system ushered in a new age of economic growth and job creation, a state-of-the-art energy transmission and distribution system may offer similar economic opportunities.

Currently, however, there is no federal legislation for upgrading the grid akin to the 1956 Highway Aid Act, and individual states still retain the right to approve sites for transmission lines. In addition to these regulatory obstacles, there is the matter of financing the enormous costs associated with a new delivery system. While the $11 billion earmarked for smart grid research and development projects under the 2009 stimulus plan* is a good start, analysts expect that thirty times that amount will be needed from public and private sources by 2030 to fully modernize the existing grid.[163]

---

* The stimulus plan is more formally known as the American Recovery and Reinvestment Act of 2009.

11

Nuclear Power and Clean Coal
Bridges to a Sustainable Future

Distributed renewable electricity is undoubtedly the way of the future. The problem is, we don't know exactly when the future begins. Until vast amounts of solar, wind, geothermal, hydrogen, biofuel, hydroelectric, tidal, and/or wave capacity are available and connected to the grid, the world must look to other options. The two most viable options to bridge the gap (nuclear power and clean coal) are considered in this chapter.

NUCLEAR ENERGY

Although the generation of nuclear energy results in no greenhouse gas emissions, nuclear power is different from other forms of alternative energy in two critical respects. First, the energy produced by today's nuclear power plants is not renewable, but rather is reliant upon a relatively rare type of uranium as a fuel source. Second, it is an available, time-tested, affordable technology now theoretically capable of meeting all of our electricity needs. In other words, we know that we *can* rely upon nuclear energy while pursuing the development of less mature renewable technologies—the question is whether it is worth the risks.

Vast amounts of energy are stored within the nuclei of atoms. This energy is released when atoms are combined (fusion) or split into smaller atoms (fission). The sun's intense heat is released by the fusion of hydrogen atoms into helium atoms, while nuclear power plants utilize the fission of fuel atoms to produce heat. The most common nuclear fuel source, uranium-235, is mined and processed into pellets. The uranium pellets (each about the size of a fingertip) are stacked into fuel rods that can be adjusted to control the plant's energy output.

Fission occurs when neutrons are fired into the uranium, creating lighter particles, more neutrons, and vast amounts of heat. From that point on, the processes of a nuclear power facility are not materially different from that of coal- or natural gas–fired plants. The heat, whether produced by fission or combustion, boils water to create steam, which powers turbines connected to electricity generators.

Compared to fossil fuel–generated electricity, nuclear power is remarkably clean. There are no GHG emissions and very little air pollution of any kind. Nuclear power is also extremely efficient in terms of consumption of raw materials. The fission of a single pound of uranium will produce more energy than the combustion of a million pounds of coal,[164] and a crayon-sized supply of uranium can generate enough electricity for a family for an entire year.[165]

As of January 1, 2008, there were 104 licensed nuclear reactors in the United States (down from 109 in 1994), providing 19% of the nation's electricity. Globally, there are over 400 nuclear plants providing 17% of the world's electricity. As demonstrated by Figure 17, the use of nuclear power varies greatly from nation to nation, with France deriving almost 80% of its electricity from this source.

Notwithstanding its many advantages, however, nuclear power is not a panacea that can be seamlessly substituted for carbon-laden fossil fuels. First, in the United States, electricity generation only accounts for 40% of overall energy consumption and 20% of GHG emissions.[166] Cars, planes, trains, trucks, and tractors will never run directly on nuclear power. Although nuclear energy may someday be used to isolate hydrogen for motor vehicles (Chapter 13) or create electricity for a fleet of battery-powered or hybrid vehicles (Chapter 10), it can not immediately solve current challenges faced by the transportation sector.

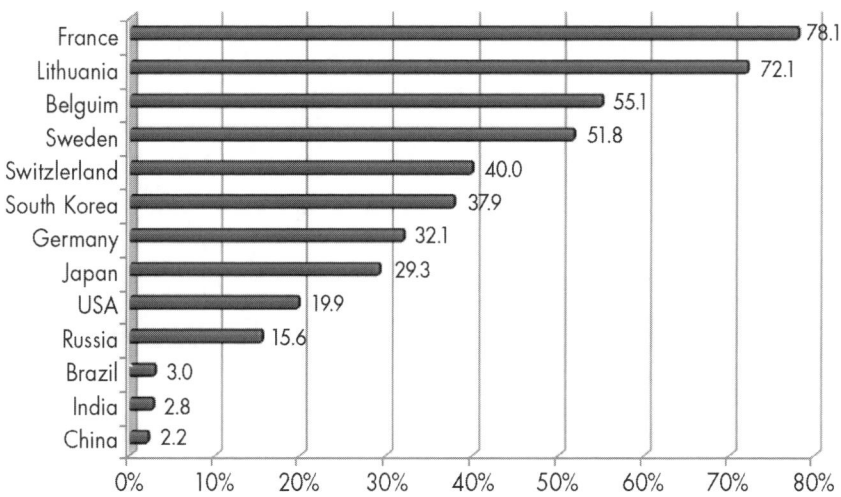

FIGURE 17. Percentage of Electricity Generated by Nuclear Power in Sample Nations

Second, the production of nuclear power plants has been limited by high start-up construction costs and health and safety concerns. In the United States, no nuclear facility has been ordered and successfully installed since the early 1970s. Yet, as cap-and-trade legislation becomes more likely, and as our vulnerability to spikes in oil costs becomes better understood, there are signs of a renaissance in nuclear development. As of October 2008, twenty-one companies sought to build thirty-four nuclear power plants in United States,[167] and new government subsidies may lead to a continuing upward trend of applications.*

While issues of high construction costs and perceived operational safety appear surmountable, several interrelated security and waste disposal issues are more daunting. A uranium fuel rod will effectively produce power for three to four years. After that, it remains dangerously radioactive well into the future. The realistic fuel rod disposal options are limited. The waste can be left on-site

---

* By way of example, on August 11, 2009, Florida approved construction of the state's first new nuclear power plant in more than three decades. The plant will be located on 5,000 acres of wetland north of Tampa.

in steel-lined pools (the current option of necessity), buried beneath the ocean floor, sent into outer space, or it can be disposed of underground in an existing geologic repository. Of these options, the National Research Council of the National Academy of Sciences, the U.S. Department of Energy (DOE), and Congress, have all chosen deep underground storage.

This selection led to the 1982 Nuclear Waste Policy Act requiring the Department of Energy to locate and develop an appropriate geologic disposal facility. After a multi-billion-dollar process, Yucca Mountain in Nevada was selected as the national repository for nuclear waste. In 2006, with the authority of Congress and president George W. Bush, the DOE proposed opening this facility on March 31, 2017.

The Yucca Mountain disposal site, which is ninety miles from Las Vegas, has been besieged with legal and political obstacles. While campaigning for the presidency in Nevada in 2008, then-Senator Barack Obama said, "There are still significant questions about whether nuclear waste can be stored safely there [and] I believe a better short-term solution is to store nuclear waste on-site at the reactors where it is produced, until we find a safe, long-term solution that is based on sound science."[168] In November 2008, Nevada senator and Senate Majority Leader Harry Reid, pronounced the project dead: "Yucca Mountain is history, O.K.? Just watch, we'll see what happens real soon, just watch. You'll see it bleed real hard in the next year."[169]

In March 2009, the Obama administration turned Senator Reid's words into political reality, announcing that the Yucca Mountain site was no longer viewed as an option for storing reactor waste and slashed its funding from the budget.[170] Thus, with little explanation and no contingency plan except "further government study," the $13.5 billion Yucca Mountain program was scuttled, and sixty thousand tons of used reactor fuel will remain indefinitely at power plants around the country.[171]

Frustration over the slow progress of the development of a geologic solution has prompted debate over a controversial method of reducing the amount of waste that must be disposed of, known as the reprocessing option. Technology exists to recycle uranium and plutonium from nuclear waste. Although the United States has barred this practice since the 1970s for security reasons, many countries including France, Russia, and India reprocess their spent fuel rods. The problem with this option is that the reprocessed plutonium waste is weapons grade, meaning that transported and stockpiled plutonium is potent enough to create a nuclear bomb. To date, worldwide commercial processing plants have been inefficient at tracking extracted plutonium—twenty pounds of which could create a devastating weapon. In a very real sense, the "miracle" of nuclear energy—the enormous stored power of the atom's nucleus—is also its curse.

Realistically, if the world is to stabilize atmospheric greenhouse gas levels and avoid a drastic reduction in its standard of living, either nuclear energy or clean coal must serve as a bridge between our current dependence on fossil fuels and the developing fleet of renewable energy options. Nuclear power is by far the more technologically advanced option of the two, but a national commitment is required to address the costs, safety (real and perceived), waste disposal, and security issues. While such outcome is theoretically achievable in the United States, there is, at present, little indication that our political leadership is inclined to make that commitment.

CLEAN COAL

Coal is the world's most abundant fossil fuel, and its use as a heating source can be traced back to Roman-occupied, second-century Britain.[172] A century ago, coal was the primary source of transportation power, used in steam-driven ships and trains. Liquefied coal effectively fueled German tanks and planes during World War II. Today, coal is used primarily to generate electricity. At most coal-fired electricity plants, the coal is pulverized into powder that, when combusted to create steam, spins turbines connected to generators. Coal-fired plants now account for 49% of the United States' electricity, and 37% of all electricity worldwide (see Figure 18). Despite growing environmental concerns and technological

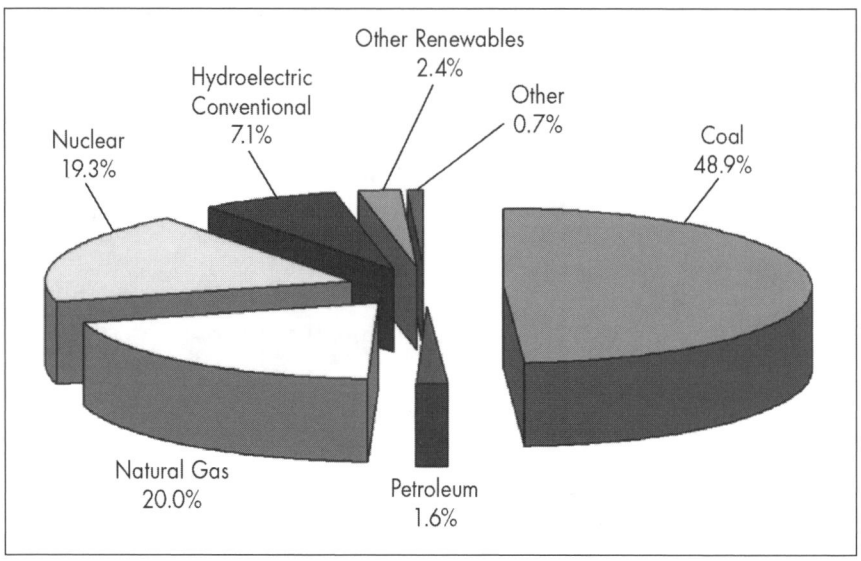

FIGURE 18. U.S. Electricity by Energy Source *(Source: U.S. Department of Energy Information Administration)*

advances in competing energy sources, the coal industry's share of total U.S. energy generation has steadily increased since 1970.

From an American perspective, there is much to like about coal. Unlike oil and natural gas production that will soon begin to decline (see Chapter 9), sufficient coal reserves exist to support current energy production levels for another 130 years.[173] Despite having a mere 6% of the world's land mass, the U.S. controls 27% of all proven recoverable coal reserves, and even with its prodigious consumption, the U.S. is a net exporter of coal. As then-Senator Obama said while interviewed on *Meet the Press:* "We're the Saudi Arabia of coal, and I don't think that we can dismiss out of hand the use of coal as part of our energy mix."[174]

As concerns climate change, however, the use of coal comes with a heavy toll. In yielding the same energy output, coal produces twice the CO_2 emissions of natural gas, and one-and-a-half times the emissions of oil. Coal combustion also discharges higher relative levels of sulfur dioxide and nitrogen oxide (which combine to cause acid rain), mercury, and carbon monoxide. Absent some economical means to clean or capture the CO_2 emissions, coal usage is ultimately irreconcilable with any meaningful international or domestic cap and trade program (see Chapter 7). As a result, companies seeking to build new coal-fired plants have experienced increasing difficulty in obtaining financing and regulatory approvals. Several major banks have adopted the "Carbon Principles," recognizing that every loan analysis should include "enhanced diligence" examining the impact that a future emissions control program is likely to have upon the financed operation.[175]

As discussed earlier, a tenuous coalition of interests exists between climate solution advocates and those concerned with U.S. dependence on foreign oil. These growing constituencies share a common motivation to support the development of solar, wind, tidal, hydro, biofuel, geothermal, wave, nuclear, and hydrogen technologies. This fragile alliance parts ways, however, when it comes to the subject of coal. To climate solution advocates, coal is the least attractive option—"an energy dinosaur whose time has passed."[176] To the energy security constituency, however, coal is America's "ace in the hole," an abundant resource capable of delivering another century of prosperity. At the heart of this intense debate is the role of clean coal in our future energy portfolio.

The controversy over clean coal begins with the term itself. Anti-coal and pro-coal advocates engage in an ongoing struggle seeking to characterize coal as "dirty" and "clean," respectively. In this battle over public perception, many simply reject use of the term "clean coal" altogether. Among those who do use the term, there is great diversity in what they mean. The most commonly accepted meaning of clean coal is the use of technology for the capture and underground storage of CO_2 emissions. To some within the coal industry, however, clean coal is defined much more broadly. For example, the American Coalition for Clean Coal Electricity, an industry trade group, uses the term to describe "any technology to reduce pollutants associated with the burning of coal that was not in widespread use" before 1990 regulations.[177] Buried deep within all the rhetoric lies a legitimate and very important question: Is there an unlocked technology that could safely and cost effectively sequester (i.e., store) CO_2 emissions below ground while we develop a viable combination of renewable energy substitutes?

Technologies exist to capture CO_2 either before or after the coal is combusted. Post-combustion capture occurs at the exhaust point, where emissions are treated ("scrubbed") with solvents. The removed CO_2 is then pressurized, liquefied, and moved to a site where it can be buried underground. Suitable subterranean locations include depleted oilfields (where injected CO_2 will enhance oil production), unminable coal seams, and saline formations (and, perhaps, the depths of the ocean*).

In pre-combustion capture (also known as "integrated gasification combined cycle" or "IGCC" because it involves multiple steps), the CO_2 is chemically removed from the coal *before* it is ignited. In step one, the coal is combined with oxygen to create a syngas. In step two, the syngas is separated into streams of hydrogen and CO_2. The CO_2 is then compressed, transported, and buried; the hydrogen can be used to operate zero-emission power plants or automobile fuel cells (see Chapter 13).

In addition to some unsolved technological challenges on the finer points, carbon capture and storage (CCS) comes with a hefty price tag. Although estimates vary greatly, it is anticipated that widespread use of CCS technology would increase the cost of coal-fired power somewhere in the range of 37 to 91%.[178] That higher cost includes a substantial increase in the amount of coal used to fuel the sequestration process. Moreover, due to the additional steps in the energy production process, CCS plants are likely to experience more time off-line than traditional coal plants. While coal CCS offers the holy grail of energy security and lower emissions, its feasibility depends on substantial public and private

---

* Ocean sequestration, as its name suggests, is storing carbon in the ocean. We discussed how oceans can be fertilized with iron to capture atmospheric carbon in the Carbon Sinks section of Chapter 2. Carbon can also be sequestered by direct injection into the oceans' depths. There is uncertainty, however, concerning how these deep sea "carbon lakes" might impact eco-systems, and how carbon injection might slow natural sequestration processes. As a result, further study must precede any widespread application of this approach.

investment as well as some form of carbon tax (or cap) which penalizes uncaptured emissions (see Chapter 8).

To date, only a few industrial-sized CCS projects have been put into operation. Norway's Sleipner natural gas field below the North Sea, operational since 1996, is the world's oldest such project. CO_2 is scrubbed from natural gas recovered from the field, and stored one thousand meters below ground. The world's first coal-fired power plant to employ CCS technology went online in September 2008 in Spremberg Germany. After a three-year initial testing phase, the 30-megawatt pilot facility (serving the equivalent of eighteen thousand homes),* will be scaled up to 300 megawatts.

In the United States, the FutureGen Project was intended to be the world's first near-zero-emissions coal-fired power plant to utilize pre-combustion CCS technology to generate electricity and hydrogen for fuel cells. After a competitive bidding process, Mattoon, Illinois, was selected by the Department of Energy as the site for the $1.8 billion IGCC project. Shortly after Mattoon was chosen, however, the DOE withdrew its funding from this flagship project due to "cost escalation." This strangely timed decision led U.S. Senator Dick Durbin of Illinois to remark: "When the city of Mattoon, Illinois, was chosen over locations in Texas, the secretary of energy set out to kill FutureGen."[179] While FutureGen mires in political stasis,† China is moving forward with its rival GreenGen project. When GreenGen goes on line in 2011, it will be China's first coal-fired power plant to employ pre-combustion IGCC to achieve near-zero-emissions.

In a comprehensive 2007 report, the Massachusetts Institute of Technology aptly summed up both the importance and the complications of developing commercially viable CCS projects.[180] Some of the key observations follow:

---

* This calculation is based on the assumption that 1 megawatt of coal power will meet the demand of six hundred homes. The megawatt to number of homes ratio is discussed further in Chapter 10.

† In July 2009, the Energy Department reached a cooperative agreement on preliminary plans for a revival of FutureGen, and a final decision is expected in 2010. www.netl.doc.gov/publications/press/2009/7637.html.

- "Coal use will increase under any scenario because it is cheap and abundant." Absent an unforeseen "technological breakthrough" in renewable energy, coal will remain indispensable.

- "CO_2 capture and sequestration (CCS) is the critical enabling technology that would reduce CO_2 emissions significantly while also allowing coal to meet the world's pressing energy needs. . . . Successful implementation of CCS will inevitably add cost for coal combustion."

- While it is critical that the government not fall into the trap of choosing a "technology winner," immediate government support is needed for building demonstration projects like Future-Gen, and for supporting research and development.

The MIT report suggests that these objectives can be met with creation of a "new quasi-government Carbon Sequestration Demonstration Corporation" to select those demonstration projects that would receive financial assistance.

Having looked at nuclear power and clean coal as potential paths to the future, we turn our attention to longer-term renewable energy solutions. Chapter 12 evaluates several types of sustainable energy that generate electricity, such as wind, solar, hydro, geothermal, tidal, and wave power. Chapter 13 considers the two renewable options most commonly associated with transportation: biofuels and hydrogen.

12

Renewable Options— Electricity
Reading the Menu at the Last Chance Saloon

To maintain any semblance of our progressively high-tech world, we will need to produce increasingly massive amounts of energy, while finding a way to substantially reduce greenhouse gas emissions. To accomplish this gargantuan task, we can choose from any combination of nine sustainable options. Two of these alternatives, nuclear power and clean coal, were considered in the last chapter. Two renewable energies that are particularly amenable to light vehicle transportation, hydrogen and biofuels, are discussed in the following chapter. This chapter looks at the pros and cons of each of the five other options on today's sustainability menu: geothermal, wind, wave (and tidal), solar, and hydropower.

GEOTHERMAL POWER

From the Earth's outside in—that is, from its crust, through its mantle, to its white-hot core—temperatures increase sharply. Just two miles below the Earth's surface, the rock is hot enough to boil water. Where the upper mantle meets the lower crust, about twenty-five miles below ground, lies a layer of super-heated, liquefied rock called magma. In certain locations, particularly where the Earth's crust is thin, or where tectonic plates come together and push magma upward, heated water can break through the surface in the forms of hot springs or geysers. The magma, and the rocks and water that are warmed by the magma, constitute a vast resource of geothermal energy.

For simplicity's sake, it is helpful to consider the three basic types of geothermal energy—"direct use" geothermal, ground source heat pump technology, and geothermal electric power plants—separately. The earliest human applications of geothermal energy all involved the direct use of heated water and steam near the surface. Since at least the last ice age, the Indian tribes of western North America used hot springs for cooking, bathing, and heat.[181] Ancient Romans used geothermal power to heat bathhouses and dwellings, and the Maori of New Zealand used heated water for cooking, washing, and warmth. Today, where available, hot water at or near the Earth's surface is transported in pipes to warm greenhouses, melt snow on roads and airport runways, and heat buildings. Indeed, in Reykjavik, Iceland, 95% of the buildings are warmed by hot water from geothermal pools.

The second type of geothermal power involves the use of a ground source (or geothermal) heat pump. Although surface temperatures vary greatly both by season and by geographic locale, in most areas the layer of ground just a few feet below the surface maintains a constant year-round temperature of between 50° and 60°F (10° and 16°C). In the winter, an electric heat pump can efficiently pull this relatively warmer air through ductwork, providing heat to buildings. In the summer, the system can be reversed, transferring heat from within the building into the adjacent ground.

Geothermal heat pumps can reduce electricity use (and associated CO_2 emissions) by up to 72%, and the operational cost savings of a typical system will exceed its installation costs within a few years.[182] By the end of 2008, there were approximately 750,000 ground source heat pumps operating in the United States, and another 50,000 are installed each year.[183]

The first two types of geothermal energy, direct use of hot water and ground source heat pumps, are limited to heating and cooling applications. The third form of geothermal energy, electricity produced by geothermal power plants, has far wider potential. The benefits of geothermal electric power are substantial. Because it requires no fuel combustion, emission levels are almost nonexist-ent. Unlike intermittent sustainable resources such as wind and solar, its capacity factor (output reliability) compares favorably to that of fossil fuels. Although there are limits to the amount of heat that can be extracted from any one location during a particular time period, exploited locations are replenished naturally. Finally, where a suitable heat reservoir is available, geothermal electricity can be very cost-effective.

Geothermal heat is converted into electric power in several ways. As discussed in previous sections, power stations can use steam to turn turbines and generate electricity. At dry steam plants, such as the 85-megawatt Geysers site in Northern California, the steam comes ready-made from the ground. Specifically, under-ground steam is pulled from a geothermal reservoir through giant strawlike pipes, then fed into the turbines. The process is a bit more complicated when magma-heated water must first be converted to steam. In this case, either a flash steam plant will change the heated water to steam directly, or a binary plant will introduce the heated groundwater to a liquid with a lower boiling point to cre-ate steam.

Despite its many advantages, the development of geothermal power has been constrained by the three critical factors: location, location, and location. In places such as Iceland (which derives 19% of its electric power from geothermal heat), the Philippines (27%), Mexico (3%), and Africa's Great Rift Valley, magma has been pushed close enough to the Earth's surface to make electricity pro-duction economically viable. The majority of the planet, however, has not been bestowed with superheated surface pools or an easi-ly accessible steam source. The perception of geothermal power as merely a localized energy source has caused it to receive consider-ably less attention than wind, solar, and hydrogen technologies, and it has been dubbed by some "the poor cousin" of the renew-able energy family.[184] In 2006, however, an MIT-led research team concluded that, with the use of a developing technology known as Enhanced Geothermal Systems (EGS), geothermal electricity could become a major player in the United States energy portfolio.[185]

EGS technology, also known as Hot Dry Rock (HDR) technology, involves pumping highly pressurized water through a shaft into fractured rocks miles below the Earth's surface. The scorched water is then recovered through a second borehole and used to generate electricity through a flash steam or a binary process. Once cooled, the water is re-injected into the ground as part of a closed-loop system. In October 2007, the world's first commercial EGS pilot project went on line in Landau, Germany, producing a constant flow of water heated to 302°F (150°C). Although the Landau plant produces a modest 3 megawatts of electricity, substantially larger projects are being developed in Australia and the United States.

The MIT study summarized the long-term prospects of EGS as a commercially viable energy source as follows:

> [B]ased on our technical and economic analysis, a reasonable investment in R&D and a proactive level of deployment in the next 10 years could make EGS a major player in supplying 10% of U.S. baseload electricity by 2050. . . . The potential of EGS in evolving U.S. energy markets is large and warrants a comprehensive research and demonstration effort to move this technology to commercial viability, especially as the country approaches a period when the gap between demand for and generation of electricity will most affect the existing system capacity.[186]

With recent increases in EGS investment by companies such as Google, and with $400 million directed to geothermal applications under the 2009 stimulus bill,[187] it is growing increasingly likely that the "poor cousin" will not be poor forever.

HARNESSING THE WIND

*"When the wind rises, some people build walls.
Others build windmills."*
—CHINESE PROVERB

Wind is an abundant, renewable form of kinetic energy generated by differential temperature around the world. Because the equator receives more direct sunlight than the poles, its air is consistently warmer. As the heated equatorial air rises, the lower atmosphere becomes less dense, and cooler air rushes down from the poles to replace it. This transfer of air from a high-pressure area to a low-pressure area, which is also influenced by the Earth's rotation, creates wind. Electricity is generated from wind in much the same way as it is generated from the kinetic energy of rivers. The wind impacts turbine blades, causing the rotation of a shaft connected to a generator.

Although modern commercial wind turbines did not come into prominence until the 1980s, wind has been used to sail watercraft, irrigate crops, and grind agriculture for centuries. In 2008, wind plants generated just over 1.5% of the world's electricity,[188] or about the same amount as eight large nuclear power plants.[189] That unspectacular statistic, however, fails to tell the full story.

The twenty-first century has seen exponential growth in wind-generated power, both in the United States and elsewhere. Worldwide wind-power capacity quintupled between 2000 and 2007, and in the United States, capacity has grown at an annual rate of 30% between 2003 and 2007.[190] In 2008 alone, the U.S. increased its wind capacity by another 50%. The World Wind Energy Association, in its 2008 year-end report, offered the following enthusiastic assessment: "Carefully calculating and taking into account some insecurity factors, wind energy will be able to contribute in the year 2020 at least 12% of global electricity consumption. By the year 2020, at least 1,500,000 megawatts can be expected to be installed globally."[191]

In 2008, the United States overtook Germany as the world leader in installed wind-generated electric capacity (Figure 19). Nevertheless, wind still supplies just over 1% of total U.S. electricity demand —a percentage that lags behind several European nations such as Holland (19%), Spain (9%), Portugal (9%), Germany (6%) and the Republic of Ireland (6%). In 2006, Texas surpassed California to become the leading domestic producer of wind power, and Texas wind capacity alone is now greater than that of all but five nations.

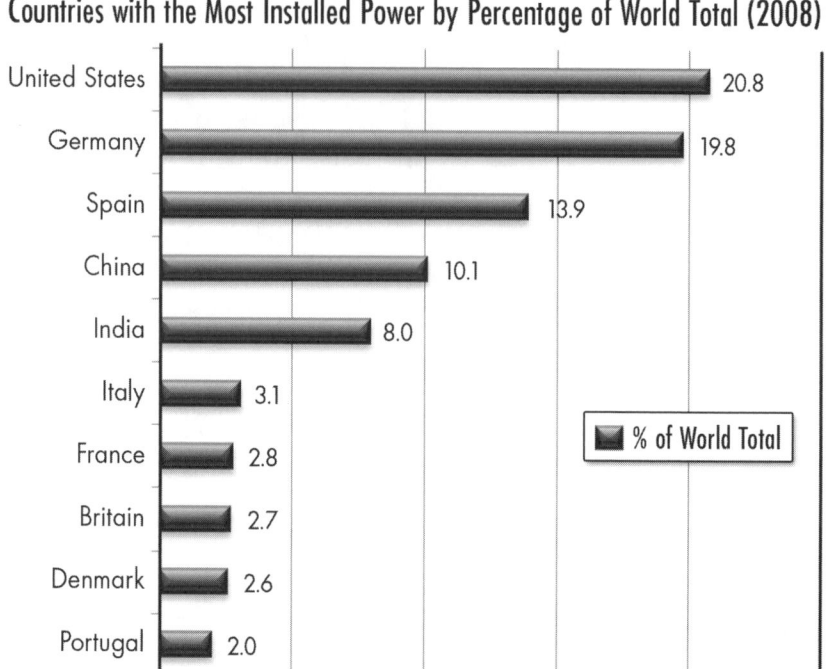

Countries with the Most Installed Power by Percentage of World Total (2008)

Country	% of World Total
United States	20.8
Germany	19.8
Spain	13.9
China	10.1
India	8.0
Italy	3.1
France	2.8
Britain	2.7
Denmark	2.6
Portugal	2.0

FIGURE 19. World Wind Power *(Source: Global Wind Energy Council)*

The North American plains, because of their level ground and exceptional wind speed, have been dubbed "the Saudi Arabia of wind." Although the entire corridor from Northern Texas to Canada is considered wind rich, areas of North Dakota, South Dakota, and Kansas are particularly exceptional. Wind speeds in these states regularly exceed 16 mph (7.2 meters per second), which is 50% higher than the average global wind speed of 11 mph (4.84 meters per second).[192] Because the relationship between increased wind speeds and increased production capacity is exponential, and because a doubling in wind speed results in an eight-fold increase in generated electricity, America's Great Plains represent "one of most promising continental areas for wind power in the world."[193]

Wind power is renewable, plentiful, clean, zero-carbon, dispersed throughout the world, and, in many locations, competitively priced

—but it is not without its disadvantages. One disadvantage is that wind speeds are highly variable, and thus wind-generated electricity is among the most intermittent of all energy sources. While a nuclear power plant may run at full output 90% of the time (a 90% capacity factor), the average international capacity factor for wind plants is around 20%.[194] In areas without a fully functional, geographically dispersed power grid (such as the United States; see Chapter 10), operators must rely upon a back-up power source.

Wind power's second disadvantage is the sheer physical space necessary to produce large amounts of electricity. Although figures are vigorously disputed, one consumer evaluation source estimates that in order to generate comparable levels of energy, wind turbines require one hundred times the land space of a conventional power plant.[195] The lack of available land—especially land of that magnitude—near urban-electricity demand poses substantial, unresolved transmission issues.[196]

Since the early 1980s, the cost of electricity generated at state-of-the-art wind-powered facilities has dropped over 80%—from thirty cents per kilowatt hour to five cents per kilowatt hour.[197] As a result, where available, wind power is now priced competitively with other forms of energy. As more plants are built, and as those projects become larger in scope, its relative cost is likely to drop even further.

WAVE AND TIDAL ENERGY

Anyone who has stood in the ocean surf (or has been repeatedly pummeled during misadventures on a longboard) has a visceral understanding of the raw power of waves. Wave energy is a broad term representing any number of methods used to convert this vast, renewable, and zero-carbon source of kinetic energy into electricity.

The previous section describes how the sun, by warming the equator more than the poles, creates wind. As the wind travels along the surface of the world's oceans, it creates tiny undulations in the water. Over long distances, as wind continues to push these ripples along, waves develop. Generally speaking, the more wind and the greater the distance traveled, the larger the wave becomes. In turn, the size, along with the speed and density of the wave, determines its potential as a source of exploitable energy. The most favorable locations for access to consistent, powerful waves are western coastlines adjacent to open ocean. Consequently, most of the existing pilot projects are in places such as Portugal, Scotland, and along the northwest coast of the United States.

While several technologies are now capable of exploiting the immense power of the ocean, none has yet to emerge as a cost-effective option to fossil fuels, nuclear, wind, or solar. For any particular form of wave energy to become economically viable, it must overcome two obstacles. First, the system must contend with the harsh surroundings of the sea. Weather conditions and corrosive saltwater can devastate equipment. Second, waves of varying sizes arrive from different directions and at different speeds. As a result, generating systems must be developed that can effectively adapt and operate at hourly, daily, weekly, and monthly variations.

Given all of these complications, the general rule of thumb in wave energy technology has been: the less moving parts, the better. Simplicity is the hallmark of the leading systems, three of which are discussed here:

1. *Oscillating Water Columns* are simply large ocean buoys tethered to the sea floor that contain a subsurface opening in the chamber. As each wave rolls in, the chamber fills with water, and the air becomes highly pressurized. The expelled pressurized air, as well as the air that flows back into the chamber when the wave recedes, is used to spin an electricity-generating turbine.

2. *Overlapping Devices* use V- or U-shaped tapered walls to channel waves up a ramp and into a reservoir. The captured water is fed back down to the ocean through a turbine outlet. The falling water generates electricity by striking and turning turbines blades—a process similar to that used to generate hydroelectricity (which is described later in this chapter).

3. *Attenuator Systems*, like the Pelamis Wave Energy Converter, are comprised of long cylindrical tubes which extend out from the

shore and float on the water. Each converter consists of a series of these tubes, which look like foam noodle pool toys connected at hinged joints. As waves roll in, the surge moves the sections in different directions relative to each other, creating hydraulic pressure. On September 23, 2008, a Pelamis project three miles off the coast of Portugal began sending power to the Portuguese grid—becoming the world's first commercial-scale wave farm.[198]

As compared to other renewable energies such as wind and solar, wave power is an immature technology with many unsolved problems. Even those in the growing ocean energy industry concede that widespread application of this form of power is probably a decade away.[199] In the long term, however, wave energy offers something that wind and solar do not: a highly concentrated power source. Because water is eight hundred to one thousand times denser than air, waves contain kinetic energy that dwarfs the energy stored in wind. As summed up by *Renewable Energy World:* "A characteristic of wave energy that suggests that it may [someday] be one of the lowest cost renewable energy sources is its high power density. Processes in the ocean concentrate solar and wind energy into ocean waves, making it easier and cheaper to harvest, according to the research team. Solar and wind energy sources are much more diffuse, by comparison."[200]

Tidal power is energy harvested from the daily rise and fall of the world's oceans. Twice each day, the gravitational pull of the moon and the sun moves huge amounts of water, altering ocean levels by as much as forty feet in some locations.[201] Two currently available technologies can convert these tidal flows into electricity. The first approach involves building huge dams (barrages) across river estuaries* to trap water in a reservoir or tidal lagoon at high tide. When the tide recedes, water is released through gates or tunnels to spin turbines, similar to the methods used by hydroelectric plants and the overlapping-wave-device facilities discussed earlier. This method offers a reliable source of clean energy, and

---

* An estuary is the wide mouth of a river where freshwater meets the saltwater of the sea.

barrage systems like the existing 240-megawatt facility in La Rance, France, have operated successfully for over forty years.

Despite these advantages, barrage technology is rarely cost-effective due to an overall lack of suitable locations and the high price of constructing a dam across a river basin. The World Energy Council, in its 2007 *Survey of Energy Resources,* projected that, due to high capital costs, any development in barrage and lagoon systems in the near future would likely take place in "combination with transport infrastructure" such as bridges.[202]

A second, more novel, approach is to build tidal turbines that use the tidal current in much the same way that wind turbines use the wind. Tidal stream systems are still in the early development stages, but because of their relatively low cost, several pilot projects have been commenced. While tidal stream technology is still many years behind wind power,[203] it offers the promise of two benefits that wind can not deliver: output reliability and a highly concentrated power source (due to water's increased relative density). According to the World Energy Council, tidal current systems offer particular opportunities in rural coastal and island communities.[204]

SOLAR ENERGY HAS A BRIGHT FUTURE

Solar radiation is an increasingly viable source of both heat energy and electricity, with vast future potential. Power from the sun is free, widely distributed, clean, and abundant. According to the U.S. Department of Energy, every hour enough solar radiation reaches the Earth to meet the world's energy needs for an entire year.[205] The challenge lies in finding ways to economically convert that energy to a practical resource.

There are countless existing and potential applications of active solar energy* but only three such technologies are currently in widespread use. Two of these applications (concentrated solar power and photovoltaics) generate electricity, and a third (solar

---

* Active solar power systems use some mechanical means to create electricity or control temperature. Passive solar energy design uses the sun's light and/or heat directly. Examples of passive-energy design include south-facing windows, shade trees, and use of cooling breezes.

heat collection) is used to heat water and dwellings. All three technologies show great promise, and use of each has grown exponentially in recent years.

Nature has provided us with several sources of stored solar energy. As discussed previously, fossil fuels, biofuels, wind, and wave energy are each forms of concentrated solar power. Unfortunately, direct solar energy is dispersed quite inconveniently across every square inch of the planet. This natural diffusion of heat and light is one of two great impediments (the sun's intermittency being the other) to each of the leading solar energy applications discussed next.

Concentrated solar power (CSP) technology uses mirrors and lenses to channel the energy of a large sunlit area into a much smaller area occupied by a "working liquid." The superheated liquid is then used to make steam that turns a turbine and generates electricity. Most CSP systems use parabolic (U-shaped) troughs that are adjusted throughout the day to capture solar radiation. Depending upon the particular design, the mirrors will redirect the sunlight to either a central tower or a tube running through the middle of the troughs containing the working fluid. As of 2009, the Solar Energy Generating (SEGS) facility, consisting of nine plants in the Mojave Desert, was the largest CSP operation in the world. The SEGS facility, with almost one million mirrors covering 1,600 acres, generates 354 megawatts of power.

The second way of using the sun's light to produce electricity is known as photovoltaics (or simply PV). The photovoltaic effect is the direct conversion of sunlight into electricity by use of solar panels. The panels are made up of PV cells that use solar energy particles (photons) to create an electrical current. A PV cell is simply two pieces of silicon connected to each other—one positively charged and one negatively charged. When photons strike the PV cell, they free electrons which seek to flow from the negatively charged silicon layer to the positively charged layer and can be channeled into a direct electric current.*

---

* Before this power can be supplied for general use, it must be run through an inverter to convert it from direct current (DC) to alternating current (AC).

Unlike CSP projects, which are usually large and commercial in scope, photovoltaic applications can range from small calculators to single homes to large electric plants producing several megawatts.* Historically, PV cells supplied electricity to dwellings and businesses in remote off-grid, locations. Since 1999, however, a majority of installed PV applications have been connected to the grid.[206] Until recently, it was largely assumed that CSP would be the dominant grid-connected solar application for the foreseeable future. Several recent developments in PV technology have raised doubts concerning that assumption—and right now, PV is competing with CSP and with other renewable energies for a viable share of the world's evolving energy portfolio.[207]

The third solar technology already in widespread use is solar water heating. A typical solar water system utilizes a roof-top heat collector—essentially a flat, black box covered with glass—to capture the sun's rays and heat an antifreeze fluid. That heated fluid is then pumped to a heat exchanger inside the building's hot water storage tank. As a result of the sun's intermittency, solar heating systems can provide only about two-thirds of domestic hot water use and will not completely replace conventional systems. However, because water heating consumes 15% of an average home's energy use,[208] solar water collectors can result in significant energy savings. In Israel, the installation of solar hot water heaters in new homes has been mandatory since 1980. Now, over 90% of all Israeli homes have solar collectors, satisfying 4% of that nation's overall energy demand.[209] More recently, the nations of Cypress and Spain (and the state of Hawaii) instituted similar mandatory solar heating installation programs for new home construction.

Solar water heating, CSP, and PV use are each growing at a torrid pace. While the lack of transmission lines to remote desert areas still serves as major hurdle to CSP and large-scale PV applications, U.S. government assistance seems to be on the way. In addition

---

* As of May 2009, there were 50 PV facilities world-wide with capacity in excess of 10 megawatts.

to $11 billion in stimulus money set aside for new transmission power lines, Congress (as of October 2009) was working on legislation that would preempt state regulatory bodies and empower the Federal Energy Regulatory Commission (FERC) to streamline the process of connecting renewable energy generators and population centers.[210]

Despite its surging growth in recent years, solar power still accounts for only a sliver of the world's overall energy portfolio. As of 2009, less than 1% of global energy demand was generated from solar radiation.[211] Thus, while solar power presents several promising applications, it will have to overcome many obstacles, and will face intense competition, before becoming a major player in the sustainable global energy mix.

HYDROELECTRICITY: THE POWER OF FLOWING WATER

Hydroelectricity is generated by transforming the energy of flowing or falling water. It is completely renewable, low-maintenance, cost-effective, and it emits no GHGs. River dams are constructed to accumulate water, which is then funneled down a sluice, where it strikes the rotary blades of turbines connected to electric generators. The amount of the river's energy converted to electricity is proportional to the volume of the water flow and the height from which the water falls. The larger the river and the higher the dam, the greater its potential electricity output.

Hydropower is among the oldest and most dependable sources of energy. Ancient Greeks used waterwheels for milling wheat and corn. By the 1800s, the mechanical power of flowing water had become an integral part of the farming and textile industries. The first use of water current to generate electricity took place in the 1870s, when Lord William Armstrong used hydroelectricity to power his remote estate, Cragside, in northern England. The first commercially operated hydroelectric generating plant was built a few years later in 1882 in Appleton, Wisconsin, and it delivered power to a single home and two local paper mills.

Hydropower now supplies 19% of the world's electricity.[212] Canada (57%), Brazil (80%), and Norway (99%) are among the nations that derive more than half of their electricity from rivers.[213] The United States is the world's fourth-largest producer of hydroelectricity, but very few unexploited locations still exist in the U.S.

In the earlier part of the twentieth century, rivers provided just under half of the electricity consumed in the U.S., but that figure has steadily dwindled to 10%.[214] Generating over 6,800 megawatts of electricity, the Grand Coulee Dam along the Columbia River in Washington is the largest producer of hydroelectric power in the U.S. and the fourth-largest such facility in the world.[215] The largest hydroelectric plant, and the largest power facility of any kind, is the colossal Three Gorges Dam in China, which, when fully operational, will generate 22,500 megawatts of electricity.

Although the advantages of hydropower greatly outweigh the disadvantages, there are limitations and detriments. First, anytime a river is dammed, serious environmental consequences result. Reservoirs can become acidic, dams restrict the passage of sedimentary nutrients, and the migration of fish between feeding and breeding locations is impeded. Dams along two of Europe's largest rivers, the Rhine and the Seine, resulted in the local extinction of Atlantic salmon, and the Xinanjiang Dam in China reduced the number of fish species in the Qiantang River by 20%.

Second, large-scale hydro-projects require lots of water and a steep drop which, in turn, requires a very large reservoir and dam. In developed countries, most of the locations with the potential to harness large amounts of power have already been exploited or are unavailable for other reasons, and accordingly, construction of new hydroelectric projects in industrialized nations is trending down. According to the U.S. Geological Survey, however, "Untapped hydro resources are still abundant in Latin America, Central Africa, India and China."[216] As a result, almost all of the new, large-scale hydroelectric projects are in developing nations—such as the proposed massive 39,000-megawatt Grand Inga Dam along the Congo River. The Grand Inga facility would be connected to a grid serving the entire African continent and would produce a level of electricity equal to a third of that now consumed throughout all of Africa.

Another location-related detriment to hydropower is population displacement. The $30 billion Three Gorges Dam along the Yangtze River required the relocation of over one million Chinese citizens. Construction of the Ilisu Dam, which commenced in 2006, resulted in the abandonment of at least fifty-nine villages along the Tigris River in Turkey, many of which occurred at gunpoint.[217] In short, like each of the competing energy solutions, hydropower is imperfect, and its implementation requires difficult decisions and compromise.

13

Renewable Options— Transportation
Biofuels and Hydrogen

In the United States, transportation accounts for one third of all GHG emissions. Only two other nations (China and Russia) use more energy to run their entire economies than the U.S. uses solely for transportation.[218] Thus, to meet the emission reduction goals discussed in Chapter 8, we will have to fundamentally change the way we travel. We will need to produce more fuel-efficient vehicles, we will need to drive fewer miles, and we will need to find one or more substitutes for gasoline. In chapters 10 and 11, respectively, we considered electricity and natural gas as potential replacements for petroleum as the primary source of transportation-related power. This chapter evaluates two other contenders seeking to fill the big shoes of big oil: biofuels and hydrogen. As we will see, while each option was once considered to be a golden child of sustainability, both have suffered some bumps and bruises in recent years.

BIOFUELS, POLITICS, AND LIFE-CYCLE EMISSIONS

A biofuel is a form of concentrated energy derived from recently dead plants or animals. The "recently dead" part of the definition extends far beyond that applicable to a recently released movie (two months in theaters) or a recently dropped snack (subject to the ten-second rule). As concerns biofuels, "recently dead" means anything that has not been compressed into a fossil fuel over millions of years.

There are many types of bioenergy.* The most common, large-scale, bioenergy application is liquid transportation fuel, such as ethanol and biodiesel. Ethanol is simply grain alcohol that can be used as a combustible power source. It is derived from starch-based crops (such as corn and wheat) or sugar-based crops (such as sugar cane and sweet sorghum), through a fermentation process. Ethanol can also be produced (although not yet economically) from cellulose found in the fibrous, inedible parts of plants or from non-crop vegetation (such as switchgrass and miscanthus) that grows on non-tillable prairie.† By contrast, biodiesel is derived from isolating the fatty acids found both in the natural oils of plants (like algae, soybean, and sunflower oils) and in animal feedstocks. In short, ethanol is alcohol derived from sugars and starches, and biodiesels are chemically processed oils and fats derived from plants and animals.

Biofuels have much to offer. They are a form of renewable energy that exploits the natural ability of plants to capture and store solar power through photosynthesis. Because biofuels can be produced domestically in many countries, they promote energy independ-

FIGURE 20. WORLD FUEL ETHANOL PRODUCTION (2007)			
COUNTRY	MILLION GALLONS		
USA	6,499	China	486
Brazil	5,019	Canada	211
European Union	570	Other	316
		Total	13,102

(Source: Renewable Fuels Association)

---

* In addition to biofuels such as ethanol and biodiesel, bioenergy can also be derived from wood, animal waste, methane from landfills, and biodegradable waste.

† According to a 2008 USDA report, cellulose ethanol has long term potential but still requires significant additional research.

ence and foster economic growth in rural areas. Biofuels are produced in liquid form, so they can supplement (or, with some engine modifications, replace*) existing petroleum-based products.

Brazil, the world's second-largest producer of ethanol, is generally regarded as the first sustainable biofuels economy (see Figure 20).[219] As a result of commitments that the Brazilian government made shortly after the 1973 OPEC oil embargo, that nation now produces 50% of its transportation fuels from sugar cane grown on only 1% of its total arable land.[220]

Most criticism of the biofuel industry stems from the contention that cultivating crops for fuel is not the best use of the world's limited farmable land. If, in order to grow corn or sugar feedstock to make ethanol, carbon-absorbing forests need to be cleared or produce farms must be eliminated, we will have merely "robbed Peter to pay Paul."[†] Concerns about razed carbon sinks (especially rainforests) are part of the larger question of whether biofuels are ever really an eco-friendly, low-carbon substitute for oil. This increasingly heated debate centers upon a recently introduced sustainability metric known as "life-cycle" (or sometimes "wells to wheels") emissions accounting.

Life-cycle emissions accounting seeks to quantify all of the GHGs released during the growth, production, and delivery of a particular fuel. In theory, once the total direct and indirect emissions of a particular fuel are calculated, an apples-to-apples comparison with other energy sources is possible. However, attempts to measure the life-cycle emissions of U.S. corn ethanol have become a source of great controversy.

The United States is the world's largest producer of ethanol fuel, generating 6.5 million gallons in 2007. Most U.S. ethanol is derived

---

* Existing vehicles can run safely using fuel containing up to 15% ethanol (E15). Anti-corrosion modifications are required for vehicles that consistently use fuel blends with higher levels of ethanol.

† Much has been written about the relationship between ethanol, food prices, and the world's hungry. In April 2009, the U.S. Congressional Budget Office issued a report, *The Impact of Ethanol Use on Food Prices and Greenhouse-Gas Emissions*, explaining how ethanol production increases food prices and otherwise affects the economy. www.cbo.gov/ftpdocs/100xx/doc10057/04-08-Ethanol.pdf

from starches found in corn kernels. For decades, it was generally accepted wisdom that, on average, U.S. corn ethanol production was marginally carbon negative, producing about 20 to 30%[221] less life-cycle emissions than gasoline.* Corn ethanol compares unfavorably to sugar cane and sweet sorghum ethanol, which were historically understood to produce between 50 to 90% less life-cycle CO_2 than gasoline.[222]

In early 2008, however, a number of studies and articles challenged the then-accepted concept that biofuels were ever carbon negative. The gist of these publications is that previous life-cycle emission calculations failed to fully take into account crucial land use issues. According to these studies, "all biofuels resulted, directly or indirectly, in new lands being cleared," and thus, had the net effect of increasing atmospheric GHGs.[223] A February 2008 article in *Time*, "The Clean Energy Scam," sums up the purported shortcomings of earlier life-cycle-emission accounting as follows:

> There was just one flaw in the [previous life-cycle] calculation: The studies all credited fuel crops for sequestering carbon, but no one checked whether the crops would ultimately replace vegetation and soils that sucked up even more carbon. It was as if the science world assumed biofuels would be grown in parking lots. The deforestation of Indonesia has shown that not to be the case. It turns out that carbon lost when wilderness is razed overwhelms the gains from cleaner-burning fuels.[224]

On March 2, 2009, in response to a year-long onslaught of articles dismissing biofuels as decidedly non-ecological, a group of more than one hundred scientists wrote a letter criticizing what they characterized as an "unjustifiable bias" against biofuels. Nonetheless, the debate over indirect land use costs continues to intensify. Ethanol producers blame "big oil" for engaging in smear

---

* It was also generally understood, however, that in some U.S. locations, by the time the corn is grown, the starches are converted to sugar, and the sugar is converted to fuel, the ethanol production released more GHG emissions than combustion of the gasoline it replaced.

tactics, while those critical of biofuels insist that the "farm lobby" wields far too much power in Washington.

In October 2008, the Union of Concerned Scientists issued a report titled *Land Use Changes and Biofuels*. According to the UCS, while fuels made from either biomass waste or from grasses grown on land not suitable for crops have great potential, there is no reliable metric yet available to determine how the manufacture of crop-based ethanol or biodiesel ultimately impacts GHG emissions.

To summarize, great doubt remains as to whether corn ethanol reduces carbon emissions, and some uncertainty persists regarding whether even sugar and sorghum ethanol are carbon negative. The 2009 U.S. stimulus package* reflected this uncertainty, limiting loan guarantees to "leading edge biofuel projects" that will produce fuels that significantly reduce life-cycle greenhouse gas emissions. Thus, while corn ethanol will continue to be subsidized in the U.S. and will continue to reduce oil imports by approximately 3%, research will be targeted toward cost-competitive production of biofuels that do not consume broad expanses of arable land. Potential alternatives include cellulosic ethanol that can be cultivated on marginally arable land akin to prairie, biomass waste, and photosynthetic algae. In July of 2009, algae received a major boast from an unlikely source when ExxonMobil invested $600 million into the development of algae-based biodiesels.[225]

THE HYDROGEN ECONOMY

> *"Yes my friends, I believe that water will one day be employed as fuel, that hydrogen and oxygen which constitute it, used singly or together, will furnish an inexhaustible source of heat and light, of an intensity of which coal is not capable. . . . When the deposits of coal are exhausted we shall heat and warm ourselves with water. Water will be the coal of the future."*
>
> —JULES VERNE, *THE MYSTERIOUS ISLAND*

---

* More formally known as the American Recovery and Reinvestment Act of 2009.

One of the most vigorously contested issues within the larger climate change debate is the viability of a new paradigm in energy known as the "hydrogen economy." According to hydrogen energy proponents, hydrogen is a non-polluting "forever fuel" with the "potential to remake civilization along radical new lines."[226] According to its detractors, hydrogen technology is both "a cynical hoax being perpetrated . . . on the residents of planet Earth" and an "ever-unready technology" that will not be deployable for mainstream use for decades.[227] Beyond the divisive hyperbole, however, lies a nearly universal consensus of opinion: Hydrogen holds enormous potential in the long run, but its prospects for large-scale use before mid-century are limited.

It is easy to understand why hydrogen is viewed by many as the ideal substitute for fossil fuels; hydrogen atoms are the most abundant in the world, and they are highly combustible. When consumed for power, hydrogen emits no GHGs. Hydrogen fuel technology is well-established and has been relied upon by NASA for decades to power its spacecraft. The problem with hydrogen is its cost. Presently, it is simply too expensive, in terms of both dollars and life-cycle carbon emissions, to isolate, store, and deliver hydrogen to consumers.

The high cost of this fuel stems in large part from the fact that hydrogen does not exist as a freestanding element in the Earth's atmosphere. Instead, it must be extracted from an existing natural compound, such as water or natural gas. There are two commonly used processes to make hydrogen: steam reformation and electrolysis.*

Steam reformation involves the application of high temperatures to methane (CH_4) to separate the hydrogen atoms from the carbon atoms. This process is not particularly eco-friendly because: (1) fossil fuels are often used to generate the required heat and (2) the disjoined carbon is usually discharged into the air. Electrol-

---

* Because hydrogen requires an input of energy from another source to separate it into a useful fuel, it is categorized as an "energy carrier" rather than an "energy source." Energy carriers, similar to electricity, transfer energy from a primary energy source (e.g., coal, wind, or nuclear) to the consumer.

ysis is the more sustainable, but also the more expensive, of the two techniques. It uses electricity to split water (H_2O) into hydrogen and oxygen. If the electric power is generated by a renewable source, electrolysis produces hydrogen without any carbon emissions. Because of the relative expense of electrolysis, however, 95% of U.S. hydrogen is generated by steam reformation.[228]

Once the hydrogen is successfully isolated and delivered, it can be used to power automobiles in one of two ways. The simpler approach, internal combustion, burns hydrogen in much the same manner that gasoline is burned in auto engines. This approach requires only modest design modifications, and several working prototypes have already been introduced. The second option uses fuel cell technology to convert hydrogen into electricity that powers the vehicle. Through a chemical process, fuel cells are able to harness electricity that is created when oxygen and hydrogen are combined. Just as it takes energy to separate hydrogen from oxygen, energy is generated when these atoms are rejoined. In essence, the fuel cell process is the opposite of electrolysis, recapturing the energy that was used to separate hydrogen from oxygen in the first place.

Despite its exciting promise, hydrogen will not be a competitively priced transportation fuel for many years. Setting aside the high economic and energy costs of separating hydrogen from other elements, hydrogen is very difficult to deliver and store. Pound for pound, hydrogen has the highest energy content of any known element, but it is also the lightest element on Earth. Hydrogen's low density creates two major problems. First, notwithstanding its impressive energy content by weight, hydrogen stores only modest energy content by volume. For practical purposes, this means that it takes up a lot of space in transport (requiring more tankers) and in cars (requiring larger fuel tanks). Second, hydrogen's low density makes it a brilliant escape artist. When it can't break free, it diffuses into metals, weakening tanks, pipes, and valves. As a result, unless hydrogen can be processed locally, a new infrastructure of pipelines constructed with special alloys will be required.

As part of the Energy Policy Act of 2005, Congress asked the National Research Council (NRC), the functioning arm of the National Academy of Sciences, to study the maximum practical number of hydrogen fuel cell vehicles that could be deployed by 2020 in the United States. According to the NRC's 2008 report, under a best-case scenario, only 2 million of the nation's 280 million light-duty vehicles (or roughly 1 in 140) could be powered by hydrogen fuel cells by that date. The NRC also took note, however, that hydrogen may be a better long-term prospect than biofuels or electric vehicles:

> Although hydrogen could not replace much gasoline before 2025, the 25 years after that would see a dramatic decline in the use of gasoline. . . . [A]lternatives such as improved fuel economy for conventional vehicles, increased penetration of hybrid vehicles, and biomass-derived fuels could deliver significantly greater reductions in U.S. oil use and CO_2 emissions than could use of [hydrogen vehicles] over the next two decades, but . . . hydrogen offers greater longer-term potential.[229]

The NRC's conclusions about hydrogen's limited role over the next two decades represent the consensus view. Nevertheless, it is generally accepted that either a series of technological breakthroughs or increasing spikes in oil prices could accelerate the introduction of hydrogen as a common fuel.*

With neither biofuels nor hydrogen distinguishing itself as a ready solution to an increasingly immediate transportation problem, liquid natural gas and electricity have become viable options by default. How these four transportation options can best be utilized is among the concluding thoughts offered for consideration in following chapter.

---

* One such possible breakthrough involves coal gasification technology of the type employed at the FutureGen site and discussed in Chapter 11. In this process, coal is mixed with oxygen to create hydrogen and carbon dioxide that is sequestered.

14

Picking the Least-Flawed Alternatives
Adventures in the Abyss

Even as wicked problems go, climate change is pretty bad. First, there are varying opinions concerning the threat presented by global warming, with a significant minority remaining unconvinced that we have a problem at all. Second, nations will be affected differently, bear different historic responsibility, and assign different levels of priority to climate change. As a result, any international solution will require an apples-to-oranges compromise that will present formidable political barriers. Third, the solutions we seek present a moving target. New climate data, scientific breakthroughs, technological deployment, fossil fuel depletion, and a dozen other evolving factors will continue to redefine the challenge. Any successful response must be built layer upon layer, through a trial-and-error process, and will require decades of flexibility and fortitude.

Fortunately, humanity does not come upon this immense challenge with an empty arsenal. A growing army of scientists, from a variety of disciplines, is being mobilized to expand our understanding of climate change. Recent economic globalization has spurred new levels of international cooperation and has opened robust channels for communication and technology sharing. And although no single renewable energy alternative constitutes a silver bullet, promising new technologies continue to emerge.

Having spent thirteen chapters gauging the problem and inventorying the potential solutions, we conclude by examining which combination of response strategies makes the most sense.

MEASURING THE THREAT

There is no guarantee that if global GHG emissions were halved tomorrow we would avoid catastrophic climate change. Conversely, there is no way to be sure that "business as usual" for another two decades would do more harm than good. The most any given mitigation option can offer is a hedge against some unknown level of future danger. To determine how much of today's sacrifice is warranted to secure tomorrow's stable climate, we must develop some working idea of the extent of the threat.

As set forth in Chapters 1 and 6, there are good reasons to harbor doubts about aspects of the consensus view on global warming. Foremost among these reasons for skepticism is that our current perspective of the Earth's climate remains riddled with blind spots and obstructed views. Legitimate questions persist concerning how solar variance (sunspot activity), global dimming, feedback loops, urban heat islands, and several other factors influence global average temperatures. Those who caution against blind adherence to climate models, who remind us that the climate is always in flux, and who suggest that the benefits of a warmer world should be included in any comprehensive analysis, make undeniably valid points.

It is critical to recognize, however, that these well-founded reasons to keep an open mind do not change three fundamental realities. First, we know that the greenhouse phenomenon is a powerful force, without which life on Earth would not be possible. Second, we know that atmospheric levels of certain GHGs, particularly carbon dioxide and methane, have increased by as much as 40% since the dawn of the industrial revolution. Third, we know that during the last century, temperatures have risen 1°F (0.6°C) with most of that increase occurring in the last few decades. Given the fact that we understand *with virtual certainty* that ambient carbon dioxide and methane trap heat, that recent industrial activity has substantially increased ambient carbon and methane, and that global average temperatures have trended upward, it would seem spectacularly irresponsible to do nothing until a clearer picture emerges.

However, just as acknowledging some level of uncertainty should not preclude us from taking any action against the causes of climate change, acknowledging some level of risk does not justify unlimited "corrective" action. The world is full of unsolved problems, and it is important to remember that resources earmarked for climate mitigation are resources that become unavailable to address other pressing needs. As global warming commentator Bjorn Lomborg wrote in opposition to a comprehensive, global cap and trade proposal:

> There are plenty of other major global problems that have reasonably cheap solutions. One billion people lack clean drinking water. Two billion lack sanitation. Three billion lack basic micronutrients. One quarter of all deaths each year are caused by infectious diseases that we could easily combat. . . . But the atmosphere has become one in which the "good guys" fight for more money for global warming against foes real and imagined. And this polarization stops us from seeing that we need to tackle climate change the same way that we tackle most public-policy problems—by weighing benefits and costs.[230]

While many scientists take issue with Lomborg's message, his point about the prudence of performing a cost benefit analysis is unassailable. With that premise in mind, we conclude our journey by evaluating which of the available responses to climate change strike the right balance between the physical threats of inaction and the economic albatross of overkill.

PICKING THE LOW-HANGING FRUIT

Over the course of the next few decades, the nations of the world will each develop energy generation and consumption strategies.

In many countries, particularly in the United States, the negotiations that lead to these plans will be complex and contentious. There are, however, several effective and relatively painless responses to the threat of global warming that make undeniable sense and should be implemented immediately. Several of these climate change no-brainers are identified here.

Conservation

On the energy supply side, the reduction of GHG emissions will mean replacing fossil fuels with low carbon alternatives. On the demand side, however, emissions can be reduced either by conservation or by energy efficiency. Conservation involves changing behavior to save energy, while efficiency employs technology to reduce energy use without sacrificing comfort or productivity. Turning off lights and computers, recycling, carpooling, building smaller homes, and adjusting the thermostat are all examples of conservation. In his famous April 1977 speech, delivered to the nation while wearing a cardigan sweater, President Carter recognized that "conservation is the quickest, cheapest, most practical source of energy."[231]

Energy Efficiency

Energy efficiency opportunities include such diverse efforts as using architecture to take advantage of shade or the sun (passive solar); insulating buildings; and utilizing teleconferences, energy-rated appliances, fluorescent light bulbs, fully inflated tires, on - line newspapers, fuel-efficient vehicles, and white-colored roofs. Despite an abundance of catchy names—like "the fifth fuel," "negawatts," and "the invisible power plant"—energy efficiency receives far less attention than other sustainable solutions. In many cases, however, efficiency is a much simpler, and a much more cost-effective, option than other flashier renewable energy alternatives. As recognized by economist James Barrett, "[O]ver the last generation, energy efficiency has supplied three times more [new] energy for final use than all other energy forms combined. . . . We can't smell it; we can't spill it. But it's been the most important factor for growth in the last generation and will be the most important in the next generation."[232] Although energy efficiency lacks the lobbying power of fossil fuel and the cachet of wind and solar, it will necessarily be a vital part of any successful climate change strategy. As a result, it may make great sense to establish policies that will exploit this underappreciated and underdeveloped option.*

Synergizing Climate Change and Energy Security Strategies

Over the next quarter century, the United States will face two epic risks: global warming and peak oil. Each threat individually is capable of inflicting great misery, but combined, they are capable of wreaking untold calamity. We are fortunate, however, that many of the available options to mitigate the risks of climate change also simultaneously provide energy security benefits. Increased energy efficiency; power derived from wind, solar, nuclear, wind, wave, hydrogen, and rivers; and an improved power grid, are among the solutions that mitigate both problems. At a time when Americans will be asked to make difficult sacrifices, the optimum solutions are those that hedge against the risks of climate change and peak oil.

Embracing the Fact that Uncertainty Is Part of the Challenge

Perhaps because humans are hardwired to seek certainty, there is a tendency among scientists, public officials, and the media to present climate change in absolute terms. However, in the current context, that certainty is an illusion. Even assuming the consensus

---

* *Time*'s Michael Grunwald wrote an excellent article in December 2008 titled "America's Untapped Energy Resource: Boosting Efficiency." Grunwald makes several concrete proposals outlining how governments can effectively stimulate energy efficiency technology, including fuel standards, decoupling utility profits from sales volume, tax incentives, and investment into energy-saving infrastructure.

view on man-made global warming is correct, we simply do not yet know what the climate is going to do next or how different levels of anthropogenic GHG emissions might lead to different results. In the coming years, the climate picture is likely to unfold in ways that we cannot now anticipate. Any successful response to such a dynamic and unpredictable problem will require cooperation among nations, a joint strategy that remains binding but flexible, and an open-mindedness to non-consensus views. That sort of adaptive approach will only be possible if we accept that a big part of the global warming challenge lies in developing a better understanding of the nature and extent of the problem.

Internalizing External Energy Costs

The external costs of a transaction are those costs incurred by someone other than the parties involved in the transaction. External costs, or "externalities," distort the true price of commodities because they are not borne by the consumer, but rather are incurred by society at large. Examples of externalities in the energy production industry include pollution, regulatory costs, ecosystem degradation, military deployment to secure energy resources, resource depletion, land use changes, aesthetics, and increased atmospheric GHGs. Although establishing a monetary equivalent of external costs often requires subjective value judgments, some general measurements and basic observations are possible. According to ExternE, a leading European authority on the externalities of electricity generation, wind and nuclear power have the lowest external costs and coal has the highest.[233] The ExternE study also concluded that, while the externalities of wind, solar (PV), and nuclear energy generation were less than one cent per kilowatt hour, the external costs of coal were often higher than the actual price of coal-fired electricity.[234] To develop an accurate picture of what energy really costs (both internally and externally), and thus put society in a position to make sound policy decisions, the concept of externalities must work its way from the fringes of the energy discussion into the heart of the debate.

More of a Good Thing

In many locations, power generated from wind, solar, hydro, and geothermal sources is already cost-effective. Power generated from each source is zero carbon,* renewable, and results in minimal external costs. The United States, in particular, has vast untapped wind and solar resources. As the technologies for exploiting these sustainable energies improve, as economies of scale develop, and as fossil fuel supplies wane, these renewables will become even more economical. Given their environmental, climate stability, and energy security benefits, policies promoting the development of wind, geothermal, and solar (as well as hydroelectric where capacity is still available) make eminent sense.

Transmission Lines and Smart Grid Technologies

Unlike traditional power plants built near population centers, new renewable energy facilities will be built upon deserts, windy plains, and seismic hot spots. New infrastructure in the form of extra-high-voltage distribution lines will be required to move the newly available electricity to homes and businesses. Similarly, new interactive technologies, such as two-way power transmission, vehicle-to-grid connect-ability, and real time electricity pricing, have the capacity to revolutionize how power is used, generated, stored, and priced. Although these changes will require a very substantial investment, and although development of transmission lines in the U.S. will result in a greater exercise of federal authority,[235] these steps are necessary if we are serious about weaning the economy off of GHG-emitting fossil fuels.

MAKING TOUGH CHOICES

Taking advantage of the easy solutions will dramatically slow the rise in atmospheric GHGs. These solutions alone, however, are

---

* "Zero carbon" refers only to the process of converting energy to power. Construction of the plant itself, operation of on-site vehicles, and other ancillary functions ensure that no power generation facility yet built is 100% GHG free.

unlikely to resolve the climate challenges we face. Instead, governments will have to make some very difficult choices—choices that involve balancing known hardship and unknown risk; allocating limited resources; and considering issues of national sovereignty, inter-generational fairness, and historic responsibility. How the nations of the world answer these excruciatingly complex questions may well seal the fate of humanity for the next century, and perhaps beyond.

Transforming Transportation

Whether it be due to dwindling oil supplies or concerns over GHG emissions, the days of relying upon gasoline as the primary source of transportation energy are quickly coming to an end. While it is becoming increasingly clear that a viable substitute for oil is required, however, there is no consensus on what option or combination of options might fill these big shoes. Hydrogen and biofuels hold great promise, but for different reasons each present challenges that may take decades to overcome. In the shorter term, the most plausible solution seems to be an evolution from gasoline vehicles to gasoline-battery hybrids with vehicle-to-grid (plug-in) capability to electric-only vehicles. As cars and light-duty trucks are converted from gasoline to electric power, it will also be feasible to run heavy-duty trucks and buses on compressed natural gas. While these approaches squarely address the issue of dwindling oil supply, they will only reduce GHG emissions if the electricity that replaces gasoline is generated by low-carbon processes. Even with growing contributions from wind, wave, solar, hydro, geothermal, and biomass energy, you can't get there from here without either expanding the role of nuclear energy or utilizing clean coal technology.

Nuclear Power

Because of the Chernobyl disaster, the narrow escape at Three Mile Island, and terrorism concerns, nuclear power bears a heavy stigma. Nuclear energy also suffers from high initial capital costs, local

NIMBY opposition,* and a serious waste-disposal quandary. Yet, unlike other sustainable solutions, nuclear power offers the attractive combination of immediate, dependable, zero-carbon, domestic, cost-competitive energy. While nuclear energy certainly has its flaws, it is among the available technologies best suited to simultaneously address the dual threats of climate change and peak oil.

Clean Coal

As concluded in the 2007 MIT report on carbon capture and storage discussed in Chapter 11, "Coal use will increase under any scenario because it is cheap and abundant." Coal is a particularly important energy resource in the United States because the country controls 27% of all proven global reserves. But the combustion of coal is a dirty business, emitting high levels of CO_2 and other atmospheric contaminants. In theory, the process of carbon sequestration makes coal both plentiful and low-carbon. Unfortunately, however, the process remains very expensive. For carbon capture and storage (CCS) to become a viable technology, the cost of emitting carbon must increase and the cost of sequestration must decrease. While the latter will require sizable investment into CCS technology, the former will mean putting a price on CO_2.

Cap and Trade Versus Carbon Tax

The Kyoto Protocol, the climate agreement currently being pursued by the United Nations, and the 2009 U.S. Climate Bill are all cap-and-trade programs. If either the international or U.S. cap and trade negotiations fail, and there is good reason to believe at least one will, there will be renewed debate on the merits of a carbon tax. While each approach has its own merits, a carbon tax would be less expensive to administer, easier to understand, and broader in application.

---

* NIMBY is an acronym for "not in my backyard," and along with BANANA ("build absolutely nothing anytime near anything"), is used, often pejoratively to describe local opposition to new development.

Defining the Respective Roles of Governments and Free Markets in Reducing GHG Emissions

Most climate stabilization proposals agree upon the need for both heavy investment into energy research and development, and reduced barriers to bringing sustainable products to market. There is vigorous disagreement, however, over how governments and capital markets should interact to best achieve these objectives. In the United States, it is often suggested that the way out of the current energy and climate predicament is "a great mobilization of scientific and engineering brains and resources, a la the atomic bomb building Manhattan Project or the Apollo moon landing."[236] Unlike the concrete objectives behind these historic government projects, climate change presents a murky, ever-changing array of economic and political challenges. As a result, while a Manhattan- or Apollo-level commitment may be warranted, the power, wealth, and flexibility of the free market must also be part of the solution. One potentially effective approach would involve:

- Government subsidies for a broad range of promising technologies (as opposed to allowing the political process to "pick winners")

- Government assistance in building new power transmission infrastructure to even the playing field for the new generation of renewable energies

- Imposing a carbon tax, or adopting a cap and trade mechanism, to account for the external costs of fossil fuels

- Encouraging public and private prizes for important technological breakthroughs

PARTING THOUGHTS

Like other wicked problems, climate change can be solved only through adaptability, innovation, collaboration, and commitment. We don't yet know the severity of the problem, when and how it will manifest itself, what level of mitigation will be required, or how the world will share the burden. Instead of traveling a well-lit path marked with signs reading "sustainability this way," we will have to wind our way through a perilous maze of crises, changing circumstances, and incomplete information. To make matters even more interesting, this unprecedented global cooperation must occur against a backdrop of increasingly intense international competition for dwindling resources and exploding third world populations.

Will the world pull together, succeed in avoiding dangerous climate change, and enter a golden age of international cooperation? Will we fail, plunging the world into unspeakable hardship and calamity? Or, will the results fall somewhere in between? For the next few decades, the answers to these questions will play out before us. The most that can be said with any certainty today is that the chances of a positive result increase each time one of us learns more about the nature of the challenges that lie ahead.

As William Jennings Bryan recognized a century ago, "Destiny is not a matter of chance, it is a matter of choice." At this moment in history, we hold in our hands, not just our own fate, but the fate of all generations that may follow. But, unlike difficult choices made by our fore-bearers, it will not be enough to stare into the face of adversity and find the moral courage to act. Instead, our two-fold challenge lies in making prudent sustainability decisions with frustratingly incomplete information, while simultaneously pursuing a vigorous quest for knowledge. One thing is certain: passing the test will require a level of boldness, adaptability, and intellectual honesty, not found at either extreme of the climate debate.

Recommended Reading

Cool It: The Skeptical Environmentalist's Guide to Global Warming, by Bjorn Lomborg

Field Notes from a Catastrophe: Man, Nature, and Climate Change, by Elizabeth Kolbert

The Hot Topic: What We Can Do About Global Warming, by Gabrielle Walker and David King

The Weather Makers: How Man Is Changing the Climate and What It Means for Life on Earth, by Tim Flannery

Unstoppable Global Warming: Every 1,500 Years, by S. Fred Singer

With Speed and Violence: Why Scientists Fear Tipping Points in Climate Change, by Fred Pearce

Related Topics of Interest

Collapse: How Societies Choose to Fail or Succeed, by Jared Diamond

Hard Green: Saving the Environment from the Environmentalists: A Conservative Manifesto, by Peter Huber

Hot, Flat, and Crowded: Why We Need a Green Revolution—and How It Can Renew America, by Thomas L Friedman

Krakatoa: The Day the World Exploded: August 27, 1883, by Simon Winchester

Snowball Earth: The Story of a Maverick Scientist and His Theory of the Global Catastrophe That Spawned Life As We Know It, by Gabrielle Walker

The End of Oil: On the Edge of a Perilous New World, by Paul Roberts

The Long Emergency: Surviving the End of Oil, Climate Change, and Other Converging Catastrophes of the Twenty-First Century, by James Howard Kunstler

Endnotes

1. "In the Trenches on Climate Change, Hostility Among Foes," Juliet Eilperin, *Washington Post*, Nov. 22, 2009.

2. "Climategate: The Final Nail in the Coffin of 'Anthropogenic Global Warming'?," James Delingpole, Nov. 20, 2009, Telegraph.Co.UK.

3. "Climate 'Czar' Says Hacked E-mails Don't Change Anything," Stephen Dinan, *The Washington Times*, Nov. 25, 2009.

4. "Pretending the Climate Email Leak Isn't a Crisis Won't Make it Go Away," George Monbiot, November 25, 2009, Telegraph.Co.UK.

5. "The Perspective of a Scientific Skeptic," written by "Ryan O," posted to *The Air Vent*, November 23, 2009, http://noconsensus.wordpress.com/2009/11/23/the-perspective-of-a-scientific-skeptic/

6. Clive Bates, "Still Not Time to Ditch Kyoto, " cited on the Open Democracy Blog, December 9, 2007.

7. Jonathan Amos, "Deep Ice Tells Long Climate Story," *BBC News*, September 4, 2006.

8. Environmental Defense Fund, "Global Warming Skeptics: A Primer, Guess Who's Funding the Global Warming Doubt Shops?" (August 28, 2007), www.edf.org/article.cfm?contentID=4870

9. Steve Hargreaves, "Exxon Linked to Climate Change Pay Out," CNNMoney.com (February 5, 2007), http://money.cnn.com/2007/02/02/news/companies/exxon_science/index.htm

10. *Ibid.*

11. John Stapleton, "Climate Change Questioned After 2008 Tipped to Be Coolest Year of the Century," *The Australian* (January 1, 2009), www.news.com.au/story/0,27574,24861809-5009760,00.html

12. National Oceanic and Atmospheric Administration, "NOAA: 2008 Global Temperature Ties as Eighth Warmest on Record" (January 14, 2009), www.noaanews.noaa.gov/stories2009/20090113_ncdcstats.html

13. Spencer Weart, "The Discovery of Global Warming" (June 2007), http://aip.org/history/ climate/ summary.htm

14. David W. Schindler, "Carbon Cycling: The Mysterious Missing Sink," 398 *Nature* 105–107 (March 11, 1999).

15. David Roberts, "Al Revere: An interview with accidental movie star Al Gore," *Grist* (May 9, 2006), www.grist.org/news/maindish/2006/05/09/roberts

16. Stephen H. Schneider, "Don't Bet All Environmental Changes Will Be Beneficial," Vol. 5 *American Physical Society News* No. 8 (Aug./Sept. 1996 ed.), http://stephenschneider.stanford.edu/Publications/PDF_Papers/APS.pdf. Dr. Schneider has subsequently made it clear that he was not countenancing scientific dishonesty in any form, but rather identifying an ethical dilemma.

17. James Hansen, *Greenhouse Effect and Global Climate Change: Hearing Before the Commission on Energy and Natural Resources*, 100th Cong., 1st Sess. 39 (S. Hrg. 100-461 Pt. 2) (June 23, 1988) (statement of Dr. James Hansen, Director, NASA Goddard Institute for Space Studies).

18. Dr. James Hansen, "The Real Deal: Usufruct & the Gorilla," August 16, 2007.

19. U.S. EPA, "Recent Climate Change—Temperature Change" (2008), http://epa.gov/climatechange/ science/recenttc.html

20. IPCC, Summary for Policymakers, "Climate Change 2007: The Physical Science Basis." (Contribution of Working Group I to the Fourth Assessment Report of the Intergovernmental Panel on Climate Change) (2007) at 10, http://ipcc-wg1.ucar.edu/wg1/Report/AR4WG1_Print_SPM.pdf

21. International Polar Foundation, "New Research Findings Predict Seasonally Ice-Free Arctic by 2015" (Dec. 12, 2008), www.sciencepoles.org/index.php?/news/ new_research_findings_predict_seasonally_icefree_arctic_by_2015/&uid=1394

22. Mark Kinver, "Arctic Ice Thickness 'Plummets,' " *BBC World News America* (October 28, 2008), http://news.bbc.co.uk/2/hi/science/nature/7692963.stm

23. "NASA Satellite Reveals Dramatic Ice Thinning," July 7, 2009, www.nasa.gov/topics/earth/features/icesat-20090707r.html

24. Jeremy van Loon, "World's Glaciers Shrink for 18th Year in Alps, Andes," Bloomberg.com (January 29, 2009), www.bloomberg.com/apps/news?pid=20601124&refer=home&sid=ajCBthQzAiU4

25. "Scientists Say Antarctic Winter Ice Is Growing," RedOrbit.com (Sept. 12, 2008), www.redorbit.com/news/science/1553271/scientists_say_antarctic_winter_ice_is_growing

26. Kenneth Chang, "Study Finds New Evidence of Warming in Antarctica," The *New York Times* (January 21, 2009), www.nytimes.com/2009/01/22/science/earth/22climate.html?partner=rss

27. Quirin Schiermeier, "Iron Seeding Creates Fleeting Carbon Sink in Southern Ocean," *Nature*, 428, 788 (Apr. 22, 2004), www.nature.com/nature/journal/v428/n6985/full/428788b.html

28. Gerald Traufetter, "Cold Carbon Sinks: Slowing Global Warming With Antarctic Iron," *Spiegel Online International* (Jan. 2, 2009), www.spiegel.de/international/world/0,1518,599213,00.html

29. Alan Zarembo, "Ocean "Sink" May Be Plugged," *Los Angeles Times* (May 18, 2007), http://articles.latimes.com/2007/may/18/science/sci-carbon18

30. CSIRO Media Release, Ref. 08/217, "Southern Ocean Resistant to Changing Winds" (Nov. 24, 2008), www.csiro.au/news/Southern-Ocean-Circulation.html

31. "Permafrost Carbon Content Double the Old Estimate," Physorg.com (Sept. 12, 2008), www.physorg.com/news140441692.html

32. "Siberia's Rapid Thaw Causes Alarm," *BBC News* (Aug. 11, 2005), http://news.bbc.co.uk/1/hi/sci/tech/4141348.stm

33. Kathryn Hansen, NASA Goddard Space Flight Center, "Water Vapor Confirmed as Major Player in Climate Change" (Nov. 17, 2008), www.nasa.gov/topics/earth/features/vapor_warming.html

34. Randy Russell, University Corporation for Atmospheric Research, "Global Warming, Clouds, and Albedo: Feedback Loops" (May 17, 2007), www.windows.ucar.edu/tour/link=/earth/climate/warming_clouds_albedo_feedback.html

35. IPCC, Summary for Policymakers, "Climate Change 2007: The Physical Science Basis." (Contribution of Working Group I to the Fourth Assessment Report of the Intergovernmental Panel on Climate Change) (2007) at 9, http://ipcc-wg1.ucar.edu/wg1/docs/WG1AR4_SPM_PlenaryApproved.pdf

36. "2008, The Coldest Year of the Century: So much for Global Warming." January 1, 2009; http://forum.behindbigbrother.com/showthread.php?p=1201192http://forum.behindbigbrother.com/showthread.php?p=1201192

37. "Global Warming Takes Another Hit—2008 Coldest Year of the Decade" (Dec. 6, 2008), http://blogchow.com/2008/12/06/global-warming-takes-another-hit-2008-coldest-year-of-the-decade/

38. William J. Cromie, "More Frequent and Severe El Niños Expected," The *Harvard University Gazette* (Sept. 24, 1998), www.hno.harvard.edu/gazette/1998/09.24/MoreFrequentand.html

39. "Are Greenhouse Gases Causing Global Warming?" *Climate Earth* (Dec. 28, 2008), http://climateearth.blogspot.com/2008/12/are-greenhouse-gases-causing-global.html

40. Gerald Stanhill & Shabtai Cohen, "Global Dimming: A Review of the Evidence for a Widespread and Significant Reduction in Global Radiation with Discussion of its Probable Causes and Possible Agricultural Consequences," *Agricultural and Forest Meteorology*, 107: 255–278 (Apr. 19, 2001); and Michael L. Roderick & Graham D. Farquhar, "The Cause of Decreased Pan Evaporation Over the Past 50 Years," *Science*, 298: 1410-1411 (Nov. 15, 2002).

41. David Adam, "Goodbye Sunshine," *The Guardian* (December 18, 2003) (quoting David Roberts), www.guardian.co.uk/science/2003/dec/18/science.research1

42. Fred Pearce, "Clearing Smog Has Led to 'Global Brightening,' " New Scientist (May 5, 2005), www.newscientist.com/article/dn7346 ; David Sington, "Why the Sun Seems to be Dimming," *BBC News* (January 13, 2005), http://news.bbc.co.uk/1/hi/sci/tech/4171591.stm

43. Richard Black, "Cleaner Air Makes Brighter Skies," *BBC News* (May 6, 2005), http://news.bbc.co.uk/1/hi/sci/tech/4520831.stm

44. David Herring, "Earth's Temperature Tracker," NASA Earth Observatory (November 5, 2007), http://earthobservatory.nasa.gov/Features/GISSTemperature/

45. Catherine Brahic, "Behind the Scenes at the IPCC: China is the New US," *New Scientist Blogs*, February 5, 2007, www.newscientist.com/blog/environment/2007/02/behind-scenes-at-ipcc-china-is-new-us.html

46. Susan Solomon, "Temperature Variability and Extremes," *Fourth Assessment Report of the Intergovernmental Panel on Climate Change* (2007), Working Group I, Section 11.3.3.3.

47. W. R. Keatinge and G. C. Donaldson, "The Impact of Global Warming on Health and Mortality: Causes of Cold-Related and Heat-Related Deaths," *Southern Medical Journal,* November 2004.

48. Bjorn Lomborg, "Global warming will save millions of lives," Telegraph.co.uk, March 13, 2009.

49. U.S. Geological Survey, January 31, 2000, http://pubs.usgs.gov/fs/fs2-00

50. Vivien Gornitz, "Sea Level Rise, After the Ice Melted and Today," January 2007, Goddard Institute for Space Studies.

51. NASA: "Warming Is Causing Greenland Ice to Melt Faster than Expected." February 21, 2008, http://news.mongabay.com/2008/0221-nasa_greenland.html.

52. Frank J. Wentz, et.al., "How Much More Rain Will Global Warming Bring?" *Science,* Vol. 317, July 13, 2007.

53. "Climate Change: Financial Risks to Federal and Private Insurers in Coming Decades Are Potentially Significant," GAO-07-285, March 16, 2007.

54. Robert Draper, "Australia's Dry Run," *National Geographic,* April 2009.

55. A. L. Westerling, et.al., "Warming and Earlier Spring Increase Western U.S. Forest Wildfire Activity," *Science,* Vol. 313. no. 5789, pp. 940–943, August 18, 2006.

56 U.S. Department of the Interior, Press Release, "Frequently Asked Questions, U.S. Fish and Wildlife Service Proposal to List Polar Bears as Threatened Species," December 27, 2006, www.doi.gov/news/06_News_ Releases/061227faq.html

57. Union of Concerned Scientists, "Arctic Climate Impact Assessment," www.ucsusa.org/global_warming/science_and_impacts/impacts/arctic-climate-impact.html

58. Juliet Eilperin, "Climate Change Affecting Species, Study Shows," *Washington Post,* December 15, 2004.

59. Paul Jay, "The beetle and the Damage Done," *CBC News,* April 23, 2008, www.cbc.ca/news/background/science/beetle.html

60. "Birds and Climate Change; Ecological Disruption in Motion," *Audubon Society,* February 2009, www.audubon.org/news/pressroom/bacc/pdfs/Birds%20and%20Climate%20Report.pdf

61. IPCC (Fourth Assessment), "Climate Change 2007 Synthesis Report," p. 33.

62. A.P. Sokolov, et. al. "Probabilistic Forecast for 21st Century Climate Based on Uncertainties in Emissions (without Policy) and Climate Parameters," MIT Joint Program on the Science and Policy of Global Change, January 2009.

63. The CNA Corporation, "National Security and the Threat of Climate Change."

64. Ban Ki Moon, "A Climate Culprit In Darfur," *Washington Post,* June 16, 2007.

65. Randy Boswell, "Study of Arctic oil and gas show bonanza for Russia," *Canwest News Service,* May 28, 2009.

66. Commander John Patch, "Cold Horizons: Arctic Maritime Security Challenges," *Proceedings Magazine,* May 2009.

67. Doug Struck, "Climate Change Drives Disease to New Territory," *Washington Post,* May 5, 2006. But compare Christine Buckley and Helen Gibbons, "Ecologists Question Effects of Climate Change on Infectious Diseases," May 2009, http://soundwaves.usgs.gov/2009/05/research5.html

68. "Under the Weather: Climate, Ecosystems, and Infectious Disease" ("2001"), The *National Academies Press*, pages 9–10.

69. "Contribution of Working Group II to the Fourth Assessment Report of the Intergovernmental Panel on Climate Change," *Summary for Policymakers*, page 12.

70. Doug Fox, *When Worlds Collide*, Conservation Magazine, January–March 2007 (Vol. 8, No. 1).

71. "Hurricanes and Climate Change," Union of Concerned Scientists, last revised September 18, 2006, www.ucsusa.org/global_warming/science_and_impacts/science/hurricanes-and-climate-change.html

72. "Climate Models Suggest Warming-Induced Wind Shear Changes Could Impact Hurricane Development, Intensity," *NOAA Magazine*, April 17, 2007.

73. "Hurricanes and Global Warming FAQs," Pew Center on Global Climate Change, www.pewclimate.org/hurricanes.cfm

74. Gabriel A. Vecchi and Brian J. Soden, "Increased Tropical Atlantic Wind Shear in Model Projections of Global Warming," *Geophysical Research Letters*, April 18, 2007; Thomas R. Knutson, et al., "Simulated Reduction in Atlantic Hurricane Frequency Under Twenty-First-Century Warming Conditions," *Nature Geoscience*, May 18, 2008.

75. Kerry Emanuel, et al., "Hurricanes and Global Warming, Results from Downscaling IPCC AR4 Simulations," March 2008; ftp://texmex.mit.edu/pub/emanuel/PAPERS/Emanuel_etal_2008.pdf

76. Jeffrey Kluger, "Kerry Emanuel," *Time*, April 30, 2006; www.time.com/time/magazine/article/0,9171,1187251,00.html. It is noteworthy that, at a time when choosing sides occurs all too often, Dr. Emanuel had the courage and integrity to follow the science wherever it led him.

77. Robert K. Merton, *Social Theory and Social Structure*, 1968.

78. Marcia Angell, "Science on Trial: The Clash of Medical Evidence and the Law in the Breast Implant Case." *Canada Free Press*, page 159.

79. Dr. Tim Ball, "The Hockey Stick Scam That Heightened Global Warming Hysteria," May 12, 2008.

80. IPCC First Assessment Report, Figure 1 (1990).

81. John L. Daly, "The Hockey Stick: A New Low in Climate Science." www.johndaly.com/hockey/hockey.htm

82. *Surface Temperature Reconstructions for the Last 2,000 Years*, NAS Executive Summary, www.nap.edu/catalog/11676.html

83. Peter Gwynne, "The Cooling World," *Newsweek*, April 28, 1975.

84. See, e.g., Walter Sullivan, "Scientists Ask Why World Climate Is Changing: Major Cooling May Be Ahead," *New York Times*, May 21, 1975; "Warning: Earth's Climate Is Changing Faster than Even Experts Expect," *Christian Science Monitor*, August 27, 1974.

85. George Will, "Let Cooler Heads Prevail," *Washington Post*, April 2, 2006.

86. U.S. National Academy of Sciences/National Research Council Report: *Understanding Climate Change: A Program for Action*, 1975, discussed in detail at www.wmconnolley.org.uk/sci/iceage/nas-1975.html

87. Doyle Rice, "Study Debunks 'Global Cooling' Concern of '70s," *USA Today*, February 20, 2008, .

88. *Astronomical Theory of Climate Change,* NOAA, last updated April 6, 2009, www.ncdc.noaa.gov/paleo/milankovitch.html

89. Tony Phillips, "Deep Solar Minimum," *Science@NASA,* April 1, 2009.

90. I.G. Usoskin, M. Schussler, S.K. Solanki, K. Mursula, "Solar Activity over the Last 1150 Years: Does it Correlate with Climate?" (2005), www.mps.mpg.de/dokumente/publikationen/solanki/c153.pdf

91. *Ibid.*

92. Peter N. Spotts, "Are Sunspots Prime Suspects in Global Warming?" *Christian Science Monitor,* September 27, 2007 (quoting solar-climate scientist Rasmus Benestad of the Norwegian Meteorological Institute).

93. S. Fred Singer & Dennis T. Avery, *Unstoppable Global Warming—Every 1,500 Years* (2007) at 3.

94. James S. Aber, "Glacier Geology of the Kansas City Vicinity," March 31, 2005, http://minnesotafuturist.pbwiki.com/f/glacialspillways.pdf.

95. *World Climate Report,* April 21, 2008, www.worldclimatereport.com/index.php/2008/04/21/little-ice-age-in-southern-south-america/

96. Contribution of Working Group I to the Fourth Assessment Report of the Intergovernmental Panel on Climate Change, Frequently Asked Questions 6.1 (2007).

97. Thomas Gale Moore, "Warmer Days and Longer Lives," www.stanford.edu/~moore/history_health.html

98. Interview of Dr. S. Fred Singer (2000), www.pbs.org/wgbh/warming/debate/singer.html

99. "Mitigating Urban Heat Islands," www.epa.gov/heatisland/resources/pdf/heatislandsrevew.pdf

100. *Ibid.*

101. "Greenhouse Warming Scorecard," (Updated 4/2/2006) www.warwickhughes.com/hoyt/scorecard.htm

102. Kevin E. Trenberth, et al., "IPCC Fourth Assessment Report," Chapter 3—Observations: Surface and Atmospheric Climate Change (2007)

103. "Urban Heat Islands and Land Use Changes," www.warwickhughes.com/hoyt/uhi.htm

104. Brian Sussman, "The Real Climate Deniers," *American Thinker,* December 16, 2008.

105. Gavin Schmidt (April 6, 2005). "Water Vapour: Feedback or Forcing?" *RealClimate.* www.realclimate.org/index.php?p=142

106. Gavin Schmidt (6 Apr 2005). "Water Vapour: Feedback or Forcing?" *RealClimate.* www.realclimate.org/index.php?p=142

107. Peter Huber, *Hard Green: Saving the Environment from the Environmentalists (A Conservative Manifesto)* (Basic Books, 2000) at XVIII.

108. David C. Bader, et al., "Climate Models: An Assessment of Strengths and Limitations," U.S. Climate Change Science Program, July 2008, p. 1.

109. David Nicholson-Lord, "Optimum Population Trust Briefing," May 2007, www.optimumpopulation.org/opt.sub.briefing.climate.population.May07.pdf

110. Bailey, Ronald, "We're Doomed Again, Paul Ehrlich Has Never Been Right. Why Does Anyone Still Listen to Him?" The *Wall Street Journal* (Opinion Archives), May 20, 2004

111. "Tikopia: A Study of Small Island Survival," *CDNN*, January 8, 2003 (source ABC) www.cdnn.info/industry/i030108a/i030108a.html

112. Penelope ReVelle, Charles ReVelle, *The Global Environment, Securing a Sustainable Future*, page 137.

113. *Ibid.* at 137–38.

114. "United Nations Population Fund Welcomes Netherlands and U.K. Action to Avert Condom Crisis," UNFPA Press Release, November 10, 2000.

115. John A. Ross, "The Futures Group International," www.prcdc.org/images/mediafile/The_Well_Being_of_Women.pdf

116. Janet Larsen, "World Population Grew by 76 Million People in 2004: 3 Million Added in the Industrial World and 73 Million in the Developing World," Earth Policy Institute, www.earth-policy.org/Indicators/Pop/2004.htm

117. "World population to peak at 9.2 billion in 2050," Mongabay.com, March 13, 2007, http://news.mongabay.com/2007/0313-population.html

118. McDaniel, Paul, "How to Understand the Demographic Transition Model," www.ehow.com/how_2243559_understand-demographic-transition-model.html"

119. Garrett Hardin, "The Tragedy of the Commons," *Science*, Vol. 162, No. 3859 (December 13, 1968), pp. 1,243–1,248

120. "EU: Earth Warming Faster," April 7, 2009, www.reuters.com/article/environmentNews/idUSTRE5363MV20090407

121. Amy L. Luers, et al., "How to Avoid Dangerous Climate Change," Union of Concerned Scientists (September 2007), www.ucsusa.org/assets/documents/global_warming/emissions-target-report.pdf

122. *Ibid.* at 1.

123. Keith Bradsher, "China to Pass U.S. in 2009 in Emissions," The *New York Times*, November 7, 2006.

124. Deborah Solomon, "Climate Change's Great Divide," The *Wall Street Journal*, September 12, 2007.

125. John M. Broder, "House Bill for a Carbon Tax to Cut Emissions Faces a Steep Climb," The *New York Times*, March 6, 2009.

126. *Ibid.*

127. "Greece Kicked Out of Kyoto," *MINA*, April 22, 2008, http://macedoniaonline.eu/content/view/1043/53/

128. Kevin A. Baumert, "Participation of Developing Countries in the International Climate Change Regime: Lessons for the Future," *George Washington International Law Review,* (2006).

129. Paul Vallely, "The Big Question: Is the Kyoto Treaty an Outdated Failure Based On the Wrong Premises?" *The Independent*, October 26, 2007.

130. "Shared Vision for Long-term Cooperative Action," *Equity Watch—Poznan Report*, www.cseindia.org/equitywatch/shared_vision.htm

131. *Ibid.*

132. "Full Climate Deal Unlikely in Copenhagen, Warns U.N.'s de Boer," *Agence France Presse,* June 10, 2009.

133. Donna Bryson, "South Africa: Rich Nations Must Pay on Climate Change," *San Francisco Chronicle,* August 4, 2009.

134. Joseph Romm, "Obama Will Never Get 67 Votes for an International Climate Treaty in the Senate," Grist.org, December 2, 2008, www.grist.org/article/obama-cant-get-a-global-climate-treaty-ratified-so-what-should-he-do-instea/

135. Michael von Bülow, "Japan Reduction Target Leaves U.N. Climate Chief Speechless," June 11, 2009, *Associated Press,* http://en.cop15.dk/news/view+news?newsid=1493. As discussed in von Bülow's article, because it uses 2005 as a baseline year, Japan's cuts would be substantially less than the EU's commitment (representing only an 8 to 9% decrease as compared to 1990 levels).

136. "Greenhouse Gas Goals for Major Nations," April 28, 2009 http://planetark.org/enviro-news/item/52638. On August 13, 2009, Australia's Senate rejected legislation that would have mandated deeper emission costs of up to 25% compared with 2000 levels by 2020.

137. Charles Digges, "Putin Signals Russia Will Sign Kyoto Protocol for WTO membership," Bellona.org, May 23, 2004.

138. "Each Country's Share of CO_2 Emissions," Union of Concerned Scientists, last revised May 13, 2009, www.ucsusa.org/global_warming/science_and_impacts/science/each-countrys-share-of-CO2.html

139. "China's per capita GDP to Hit US$3,000 by 2010," *Xinhua, China Daily,* updated April 1, 2008.

140. Joseph Romm, "What Will Make Obama a Great President," Salon.com, December 4, 2008; www.salon.com/env/feature/2008/12/04/obama_china/index.html

141. James Howard Kunstler, *The Long Emergency,* Grove Press (2005) at p.31.

142. M.K. Hubbert, "Nuclear Energy and the Fossil Fuels." Presented before the Spring Meeting of the Southern District, American Petroleum Institute, Plaza Hotel, San Antonio, Texas, March 7–9, 1956. Hubbert's prediction for peak oil in 1970 was one of two scenarios he offered (his "upperbound" or optimistic estimate). The most likely estimate called for an oil production peak in 1965.

143. Ugo Bardi, Marco Pagani, "Peak Minerals," Last Updated October 16, 2007, www.theoildrum.com/node/3086

144. "Former Head of Saudi Aramco: Oil has Peaked," *Global Public Media,* October 31, 2007, www.globalpublicmedia.com/former_head_of_saudi_aramco_oil_has_peaked

145. "Shell Energy Scenarios to 2050," www.static.shell.com/static/public/downloads/brochures/corporate_pkg/scenarios/shell_energy_scenarios_2050.pdf

146. Keith Johnson, "Peak Oil: Global Oil Production's Peaked, Analyst Says," May 4, 2009, *Wall Street Journal* Blog.

147. Michael T Clare, "It's Official—The Era of Cheap Oil Is Over," *The Nation,* June 11, 2009 (article originally appeared on the TomDispatch.com).

148. See January 22, 2008 e-mail, from Jeroen van der Veer to all Shell employees, regarding Shell Energy Scenarios, www.theoildrum.com/node/3548 (posted January 26, 2008 by Jerome A. Paris).

149. "How Natural Gas Works" Union of Concerned Scientists (2009), www.ucsusa.org/clean_energy/technology_and_impacts/energy_technologies/how-natural-gas-works.html

150. Clean-Energy.US, "About Electricity," Edited Mar. 27, 2009, www.clean-energy.us/facts/electricity.htm; Crystal Davis, "Booming Wind Energy Market Grows 27% in 2007," World Research Institute, http://earthtrends.wri.org/updates/node/277

151. Colorado School of Mines, "Potential Gas Committee Reports Unprecedented Increase in Magnitude of U.S. Natural Gas Resource Base," June 18, 2009, www.mines.edu/Potential-Gas-Committee-reports-unprecedented-increase-in-magnitude-of-U.S.-natural-gas-resource-base

152. Dave Cohen, "A Shale Gas Boom?" Association for the Study of Peak Oil and Gas, June 25, 2009, www.aspousa.org/index.php/2009/06/a-shale-gas-boom/

153. Joseph Romm, "Climate Action Game Changer, Part 1: Is There a Lot More Natural Gas than Previously Thought?" Grist.org, posted on June 4, 2009, www.grist.org/article/climate-action-game-changer-part-1-is-there-a-lot-more-natural-gas-than-pre/

154. "History of Natural Gas Vehicles," UnionGas.com, www.uniongas.com/aboutus/aboutng/ngv/ngvhistory.asp

155. "CNG Vehicles Around the World," CNGnow.com, (citing: *Gas Vehicles Report*, May 2008), www.cngnow.com/EN-US/VEHICLES/AROUNDTHEWORLD/Pages/default.aspx

156. "New Study Calls for World-Wide Reduction in Energy Consumption," ECNmag.com; www.ecnmag.com/Efficiency-Zone-2000-Watt-Society.aspx?menuid=

157. Nuclear Energy Institute, "U.S. Nuclear Power Plants Achieved Near-Record Level of Electricity Production in 2008." Press Release, February 3, 2009 .The photovoltaic solar capacity figure is derived from David W. Hafemeister's *Physics of Societal Issues*, page 317."

158. Sharon Begley, "We Can't Get There From Here," *Newsweek*, March 14, 2009

159. Chris Hogg, "China's Car Industry Overtakes US," *BBC News*, February 10, 2009.

160. Electric Advisory Committee, "Smart Grid: Enabler of the New Energy Economy," December 2008, www.oe.energy.gov/DocumentsandMedia/final-smart-grid-report.pdf

161. Jeffrey Ball, "The Matrix Overloaded: Clean Energy Will Depend on a New Smart Grid," October 24, 2008, The *Wall Street Journal* Digital Network, http://online.wsj.com/article/SB122479941639164453.html

162. Martin LaMonica, "Will Anyone Pay for the 'Smart' Power Grid?" May 17, 2007, CNET News.com, www.zdnetasia.com/news/software/0,39044164,62013638,00.htm

163. Democratic Policy Committee, "The Case for a 21st Century Electricity Transmission System," March 5, 2009, http://dpc.senate.gov/dpcdoc.cfm?doc_name=fs-111-1-34

164. "How Nuclear Power Works," Union of Concerned Scientists, www.ucsusa.org.nuclear_power/nuclear_power_technology/how/Nuclear Power Works

165. www.uraniumnorthresources.com/i/pdf/presidents_Letter_June_07.pdf

166. Energy Information Administration, "Official Energy Statistics From the U.S. Government," Brochure #: DOE/EIA-X012, Release Date: May 2008

167. Matthew L. Wald, "After 30 Slow Years, U.S. Nuclear Industry Set to Build Plants Again," *International Herald Tribune*, October 24, 2008.

168. Daren Briscoe, "Obama's Nuclear Reservations," *Newsweek*, November 22, 2008.

169. Steve Tetrealt, "Demise of Yucca Project Predicted," *Las Vegas Review Journal*, November 21, 2008.

170. Josef Herbert, "Nuclear Waste Won't Be Going to Nevada's Yucca Mountain, Obama Official Says," *Chicago Tribune*, March 6, 2009.

171. *Ibid.*

172. U.S. Department of Energy, "A Brief History of Coal Use," http://fossil. energy.gov/education/energylessons/coal/coal_history.html.

173. World Coal Institute, "Where Is Coal Found," www.worldcoal.org/pages/content/ index.asp?PageID=100. But compare, Jonathan Amos, "Climate Outcome 'Hangs on Coal,' " *BBC News*, December 18, 2008 (citing evidence of a substantially lesser reserve).

174. *Meet the Press*, May 4, 2008.

175. "The Carbon Principles," http://carbonprinciples.org

176. David Roberts, "Coal Is an Enemy of the Human Race," November 30 2006, http://gristmill.grist.org/story/2006/11/30/134051/34"

177. Jacob Leibenluft, "What the Heck Is 'Clean Coal'?" *Slate Magazine*, October 7, 2008.

178. Frédéric Beauregard-Tellier, "The Economics of Carbon Capture and Storage," Library of Parliament, March 13, 2006.

179. Bob Secter and Kristen Kridel, "U.S. May Dump FutureGen Energy Chief Tells Lawmakers Illinois Project Is on Ropes," *Chicago Tribune*, January 30, 2008.

180. "The Future of Coal," http://web.mit.edu/coal/

181. "Direct-Use Geothermal Energy," California Energy Commission, Consumer Energy Center, www.consumerenergycenter.org/renewables/geothermal/directuse.html

182. "How Geothermal Energy Works," Union of Concerned Scientists, www.ucsusa.org/ clean_energy/technology_and_impacts/energy_technologies/how-geothermal-energy-works.html

183. *Ibid.* See also "Geothermal Heat Pumps," U.S. Department of Energy, Energy Efficiency and Renewable Energy, updated December 30, 2008, http://apps1.eere.energy.gov/consumer/your_home/space_heating_cooling/index.cfm/mytopic=12640

184. http://bx.businessweek.com/renewable-energy/renewable-energys-poor-cousin-getting-richer-international/8491700285999739822-5bca16f287a4185120940f1eeb3ab82b/

185. "The Future of Geothermal Energy, Impact of Enhanced Geothermal Systems (EGS) on the United States in the 21st Century," Massachusetts Institute of Technology (2006), www1.eere.energy.gov/geothermal/pdfs/egs_toc_front.pdf

186. *Ibid* at page 1–33.

187. "Green Billions Fertilize the U.S. Economic Stimulus Package," *Environmental News Service*, February 20, 2009.

188. "World Wind Energy Report 2008," www.wwindea.org/home/images/stories/world-windenergyreport2008_s.pdf

189. http://science.howstuffworks.com/wind-power.htm

190. U.S. Department of Energy, "Wind Energy Could Produce 20 Percent of U.S. Electricity By 2030." Press Release, May 12, 2008.

191. "World Wind Energy Report 2008," page 8. www.wwindea.org/home/images/stories/worldwindenergyreport2008_s.pdf

192. Cristina L. Archer, Mark Z. Jacobson, "Evaluation of Global Wind Power," *Journal of Geophysical Research*, Vol. 110, June 30, 2005.

193. *Ibid.*

194. Bruno De Wachter, "The Capacity Factor of Wind Power," June 12, 2008, www.leonardo-energy.org/drupal/node/3214

195. "Pros and Cons of Wind Power," www.energy-consumers-edge.com/pros_and_cons_of_wind_power.html

196. Matthew L. Wald, "Wind Energy Bumps Into Grid Limits," The *New York Times*, August 26, 2008.

197. "What Are the Factors in the Cost of Electricity from Wind Turbines?" American Wind Energy Association, www.awea.org/faq/cost.html

198. Michael Burnham, "World's First Commercial Wave Project Goes Live," *Greenwire*, September 23, 2008. Unfortunately, the success of this project was relatively short-lived. In mid-November of 2008, due to a buoyancy problem, the Pelamis units were dismantled and brought back to shore, where they remain at the time of this publication.

199. Paul Davidson, "Marine Energy Can Be Forecast," *USA Today*, April 20, 2007.

200. "Economies of Scale Could Swell Ocean Energy," February 4, 2005, Renewable Energy World.com, www.renewableenergyworld.com/rea/news/article/2005/02/economies-of-scale-could-swell-ocean-energy-21969

201. "Astronomy 101 Specials: Tides," www.eg.bucknell.edu/physics/astronomy/astr101/specials/tides.html

202. Ian Bryden, "Survey of Energy Resources 2007," www.worldenergy.org/publications/survey_of_energy_resources_2007/tidal_energy/756.asp

203. "Turning Tides Power Test of Maine Coast," *Greenwire*, September 12, 2008 (crediting Jerry Harkavy of MSNBC.com.

204. Ian Bryden, "Survey of Energy Resources 2007," www.worldenergy.org/publications/survey_of_energy_resources_2007/tidal_energy/756.asp

205. U.S. Department of Energy, Solar Energy Technologies Program website, last updated March 19, 2009, www1.eere.energy.gov/solar/animations.html

206. www.iea-pvps.org, "Figure 2: Cumulative Installed PV Power in the Reporting Countries by Application."

207. Jonathan Lesser, Nicolas Puga, "PV versus Solar Thermal," *Public Utilities Fortnightly*, July 2008.

208. "How Solar Energy Works," Union of Concerned Scientists (referencing the U.S Department of Energy; www.ucsusa.org/clean_energy/technology_and_impacts/energy_technologies/how-solar-energy-works.html

209. Neal Sandler, "At the Zenith of Solar Energy," *Business Week*, March 26, 2008.

210. Stephen Power, "Reid to Seek Federal Role on Power Lines," The *Wall Street Journal*, February 23, 2009.

211. "Solar Energy," *National Geographic*, http://environment.nationalgeographic.com/environment/global-warming/solar-power-profile.html

212. Mike Magee, *Healthy Waters: What Every Health Professional Should Know About Water*, (Spencer Books, 2005), page 67.

213. "Global Hydropower Scenario," Environmental Resource Group, Limited, (2007), www.erg.com.np/hydropower_global.php

214. "Hydroelectric Power Water Use," *U.S. Geological Survey*, (last updated February 20, 2009).

215. U.S. Department of the Interior Bureau of Reclamation, "Grand Coulee Dam Statistics and Facts," (rev. 1/09).

216. *Ibid.*

217. Rebecca Wood, "Case Study: The Ilisu Dam, Turkey," www.dams.org/docs/kbase/submissions/ins194.pdf

218. David L Greene, Andeas Schafer, "Reducing Greenhouse Gas Emissions from U.S. Transportation," Pew Center on Global Climate Change, pages 6 and 12, May 2003.

219. Christine and Scott Gable with Mario R. Duran, "Ethanol in Brazil: The World's First Sustainable Biofuels Economy," About.com, http://alternativefuels.about.com/od/ethanol/a/ethanolinbrazil_2.htm

220. http://english.unica.com.br/download.asp?mmdCode=B62A5348-F792-4F46-9314-F89884B5E611.

221. United Nations Environmental Programme, Tool 14 "Alternative Fuels," www.unep.org/tnt-unep/toolkit/Actions/Tool14/index.html

222. *Ibid.*

223. Elisabeth Rosenthal, "Biofuels Deemed a Greenhouse Threat," The *New York Times,* February 8, 2008.

224. Michael Grunwald, "The Clean Energy Scam," *Time,* April 7, 2008.

225. Katie Howell, "Exxon Sinks $600 Million into Algae-Based Biofuels in Major Strategy Shift," The *New York Times,* July 14, 2009.

226. Jeremy Rifkin, "The Hydrogen Economy after Oil, Clean Energy from a Fuel-Cell-Driven Global Hydrogen Web," www.emagazine.com/view/?171

227. "The Hydrogen Economy: An Idea Whose Time Hasn't Come . . . Again," last updated July 15, 2008, www.econogics.com/en/heconomy.htm

228. Jeff Wise, "The Truth About Hydrogen," *Popular Mechanics,* November 2006.

229. National Research Council, "Transitions to Alternative Transportation Technologies—A Focus on Hydrogen" (2008), prepared by the National Research Council, The National Academies Press, page 2.

230. Bjorn Lomborg, "Mr. Gore, Your Solution to Global Warming Is Wrong," *Esquire,* July 15, 2009.

231. Jimmy Carter, "The President's Proposed Energy Policy," April 18, 1977. *Vital Speeches of the Day,* Vol. XXXXIII, No. 14, May 1, 1977, at 418–420, www.pbs.org/wgbh/amex/carter/filmmore/ps_energy.html

232. Katie Howell, "Economic Models Underestimate Energy Efficiency," *E&E News,* July 30, 2009, www.eenews.net/eenewspm/2009/07/30/1

233. ExternE. Externalities of Energy is a project financed by the European Union, www.externe.info

234. "External Costs, Research Results on Socio-Environmental Damages Due to Electricity and Transport" (2003) www.externe.info/externpr.pdf

235. Steve Gelsi, "National Grid Plan Presents Super-sized Headache," *Market Watch,* July 9, 2009, www.marketwatch.com/story/power-showdown-looms-over-new-utility-lines

236. "A Manhattan or Apollo Project for Energy? What Nonsense," *The Chronicle,* April 6, 2008, http://chronicle.com/blogPost/a-manhattan-or-apollo-project-for-energy-what-nonsense/5844

Index

About the Author

William Stewart

William Stewart has spoken on climate change at numerous national forums and his work has been featured by NBC News, *The Wall Street Journal*, and several insurance industry publications, including *Business Insurance, Best's Insurance*, and *CPCU Journals*. He has a bachelor's degree from St. Joseph's University and a law degree from Notre Dame University. Stewart has practiced law at Cozen O'Connor, a 550-attorney law firm with offices in 24 cities on two continents, since 1990. He is the head of its Climate Change/Global Warming Division.